Sitings

Hawai'i Studies on Korea

Wayne Patterson
The Ilse: First-Generation Korean Immigrants, 1903–1973

Linda S. Lewis
Laying Claim to the Memory of May: A Look Back at the 1980 Kwangju Uprising

Michael Finch
Min Yŏng-hwan: A Political Biography

Michael J. Seth
Education Fever: Society, Politics, and the Pursuit of Schooling in South Korea

Chan E. Park
Voices from the Straw Mat: Toward an Ethnography of Korean Story Singing

Andrei Lankov
Crisis in North Korea: The Failure of De-Stalinization, 1956

Hahn Moo-Sook
Young-Key Kim-Renaud, trans.
And So Flows History

Timothy R. Tangherlini and Sallie Yea, eds.
Sitings: Critical Approaches to Korean Geography

HAWAI'I STUDIES ON KOREA

Sitings

Critical Approaches to Korean Geography

edited by
Timothy R. Tangherlini
and
Sallie Yea

University of Hawai'i Press, Honolulu
and
Center for Korean Studies, University of Hawai'i

Printed in the United States of America
13 12 11 10 09 08   6 5 4 3 2 1

Library of Congress Cataloging-in-Publication Data

Sitings : critical approaches to Korean geography / edited by Timothy R. Tangherlini and Sallie Yea.
p. cm.—(Hawaii studies on Korea)
Includes bibliographical references and index.
ISBN 978-0-8248-3138-7 (hardcover : alk. paper)
1. Human geography—Korea (South). 2. Korea (South)—Social conditions. 3. Korea—Civilization. I. Tangherlini, Timothy R. II. Yea, Sallie.
GF659.S57 2008
304.2095195—dc22

2007025514

The Center for Korean Studies was established in 1972 to coordinate and develop resources for the study of Korea at the University of Hawai'i. Reflecting the diversity of the academic disciplines represented by affiliated members of the university faculty, the Center seeks especially to promote interdisciplinary and intercultural studies. Hawai'i Studies on Korea, published jointly by the Center and the University of Hawai'i Press, offers a forum for research in the social sciences and humanities pertaining to Korea and its people.

University of Hawai'i Press books are printed on acid-free paper and meet the guidelines for permanence and durability of the Council on Library Resources.

Designed by University of Hawai'i Press production staff
Printed by The Maple-Vail Book Manufacturing Group

Contents

Acknowledgments

THIS VOLUME IS the end product of a long series of conferences, workshops, symposia, meetings—both formal and informal—and collaborative editing. The initial impetus for the volume began in a chance meeting between the editors and over subsequent years evolved into an international collaborative effort. The main funding for the project was provided by a multiyear grant from the Comparative and Interdisciplinary Research on Asia (CIRA) group at UCLA's International Institute. We would particularly like to thank Shumei Shih, director of CIRA, for encouraging us in the initial stages of the project and her continuing support over the course of the grant. Initial seed money for the project was provided by a Pacific Rim Research Grant through the International Institute. Additional support for the project was provided by the Center for Korean Studies, the Department of Asian Languages and Cultures, and the Department of Geography at UCLA.

Of course, this volume is a product of far more than funding—rather, it represents the joint efforts of many scholars of both geography and Korean studies. For several days in May 2004, everyone who had been involved in the project was able to meet for a lively symposium at UCLA. We would like to thank Michael Bourdaghs, Robert Buswell, Denis Cosgrove, Michael Curry, John Duncan, Nick Entrikin, Namhee Lee, Laura Nelson, and Stefan Tanaka for acting as discussants and respondents at that event. Their comments have helped shape in no small manner this volume. Also, we would like to thank our colleagues Nancy Abelmann, Donald Baker, Carol Medlicott, Mark Nathan, Hyung-il Pai, and Hijoo Son for their contributions to the symposium. Other colleagues, including Sheila Miyoshi Jager, Choon Piew Pow, and Hong Key Yoon, all contributed to earlier discussions and workshops.

At UCLA, Linda Truong played a critical role in keeping the project on budget; her hard work and good spirit ensured the success of the project. Eileen Sir at the Center for Korean Studies helped with local arrangements, and Chris Hanscom helped make sure that the symposium proceeded smoothly. Mark Pokorsky deserves thanks for seeing that the conference venues were available and that we encountered no technological glitches. Leighanne Yu provided valuable bibliographic support during manuscript preparation, and Elli Kim provided assistance in preparing the index.

At the University of Hawai'i, we would like to thank Patricia Crosby and Karl Kim for their guidance and patience as they shepherded this collection through all the various stages and its final production. That said, we as volume editors take all responsibility for any inaccuracies that may have cropped up in the editing and production of this volume.

Timothy R. Tangherlini
Sallie Yea

1

Introduction—Constructed Places, Contested Spaces

Critical Geographies and Korea

TIMOTHY R. TANGHERLINI AND SALLIE YEA

SEEN FROM SEVEN hundred kilometers out in space, the Korean Peninsula is unremarkable. Browns, grays, and streaks of white in the north give way to slightly greener patches in the south, indicating different topographical features, while the deep blues of the ocean on all three sides confirm that it is indeed a peninsula. A satellite image tells the normal person little more than that. Instead, such an image elides hundreds, even thousands of years of human history in the area. Human interaction with the environment and historical, political, and social developments all fall away in the satellite image. Such an image proposes a geographic overview—but an overview that is purely superficial. It is a view of Korea that ignores the very specific processes that occur in space and through those processes change a rather unremarkable space into a series of overlapping, often contested dynamic places. It is this type of geographic fiction that the present volume intends to combat. Borrowing a term from the English philosopher Gilbert Ryle, Clifford Geertz (1973) speaks of "thick description," but he limits the majority of his conversation to ethnographic description. "Thick description" requires very close observation of what people do, coupled to a deep and nuanced understanding of economic, political, and social processes. We argue that, just as importantly, a truly thick description must also take into account geographic processes—processes that influence how geography is constituted and how those geographies in turn influence what people do, not just once, but time and again in an endless feedback loop. Although Koreanists have for decades shown a remarkable ability to provide thick historical, political, sociological, and ethnographic descriptions—and analyses—of phenomena, there has been a gap in regard to a critical and thick engagement with geography.

For East Asia generally and the Korean Peninsula specifically, geography has long been studied in a most traditional sense—an inevitable result of both the dominance of quantitative approaches in the discipline in the 1960s through the 1990s and a product of the era of rapid modernization and economic development in South Korea. As one of the premier geographic research institutes of Korea explains, "Partly because of the efficiency-oriented economic development policies, the nation's socio-economic activities became spatially polarised, with over-concentration occurring in certain regions and deprivation in others. The resulting regional disparities and distorted national settlement patterns stood in the way of realising the national goals of balanced development and equitable distribution of citizen's welfare" (Korean Research Institute for Human Settlements 2004). Because the economic development policies of the modern era have been largely "a-spatial" and concerned more with fulfilling "efficiency objectives," geographers have subsequently been preoccupied with issues such as industrial location policy, transportation networks, and urban form and function and how to achieve balanced spatial development and population distribution in light of this developmental agenda.

Only more recently have engagements with geographies of Korea attempted to move beyond these traditional concerns with planning, themselves driven by the push for modernization. Critical cultural, historical, and social geographies of Korea have slowly begun to emerge that engage with different questions as a consequence of the onset of modernity and globalization. These are questions about national (and subnational, regional, local, imagined) identity, symbolic spaces, imagined landscapes and territories, contested places, representations of geographical identity and belonging through place/space, and geographies of resistance and dominance. Uhn Cho, in a recent contribution to a special issue of *Korean Journal* (2004, 5), for example, proposes reading Seoul "as a cultural text in order to disclose how the meaning of 'placeness' of urban space is generated and gained." A recent volume by Jager (2003) similarly interrogates sacred "national" spaces and places to reveal the politics of symbolic sites in Korea.

This volume extends the incipient work of those attempting to move geographies of Korea to more critical projects that engage with space and place as constructed, contested, and highly political arenas. The debate still rages about whether globalization erases or confirms the significance of space and place. Yet whatever side one falls on in this debate, it is undeniable that globalization and its incumbent processes of spatial erasure and simultaneous (re)localization have, for Korea, thrown questions of the geographics of belonging, identity, and construction of community into a sharp new light, begging scholars to explore the multiple intersections between peoples' uses, meanings, imaginings, and resistances in, to, and through space and place.

Critical Geographies

In critical geographies, landscapes, place, and space are seen to be as much created by imagination as by their physical reality. Following Said's (1978/1979 and 1993) seminal work on Orientalism, geographers have begun to examine the role of imagination in the construction of places, particularly the idea that places do not necessarily have a "real" existence beyond their social and cultural construction. As Said himself suggests, the Orient is "not an inert fact of nature. It is not merely there" (1978/1979, 3). The influence of this idea on the formation of some critical geography projects is twofold:

> First, it revamps long-held notions of space and landscape as inert platforms, grids, or containers upon or within which social relations unfold, and instead treats them as fluid dynamic forces which are produced by—and in turn produce—social relations. Indeed, viewed in this way, space and society become inseparable; second, they enlarge the purview of geography to embrace spaces not only bound up with material "objective" patterns and processes, but also spaces of a metaphorical "subjective" kind that may play equally impressive roles through cultural production and ideological formation (Kasbarian 1996, 530).

Thus, the constructed nature of places leads to a redefinition of landscapes as "created by people—through their experience and engagement with the world" (Bender 1993a, 1). Space, as Sibley (1995) notes, also consequently becomes imbricated in the processes of social inclusion/exclusion and being outside/inside, so that certain groups may become positioned in "geographies of exclusion" or "geographies of domination," notions that resonate throughout the essays in this volume (see, for example, the chapters by Yea, Pettid, and Song).

Apart from this emphasis on constructed places, critical geographers have also begun to consider the ways in which places are created through meaningful interactions with them. Thus, rather than viewing places as constituted from "outside" (as in Edward Said's work), some critical geographers have also begun to question the way subordinate and marginal groups confirm, contest, and rework place and space to create "geographies of resistance" that rely on "the significance of the spaces through which counter-hegemonic politics can be articulated" (Keith 1997, 278). This understanding attempts to move beyond simple oppositional frames of power and resistance, where "'resistance' stands in implacable opposition to 'power'" (Pile 1997, 1), thus dislodging received notions of geographies of resistance in which acts of resistance are viewed simply as taking place through specific geographies, such as demonstrations on the street, or around specific entities, such as land or the nation-state. Instead, the focus of critical geographies of resistance is on the ways space and place are themselves constitutive of resistance, so dislodging presumptions that "resistance is self-evident, that geography is an inert, fixed, isotropic backdrop to the real stuff of politics and history" (Pile 1997, 4). Others have focused on the ways new virtual and actual places are created by marginal groups in order that community or resistance can be expressed. This means engaging with the question of how meanings of places can be mobilized by those who live in them. As Routledge mentions, "bell hooks (1990) refers to these spaces as 'homeplaces' which act as sources of self-dignity and agency, sites of solidarity in which, and from which, resistance can be organized and conceptualized" (Routledge 1997, 361).

Another theme to emerge within the critical geography project is the way museums—and more recently, folk villages and cultural theme parks—have become important sites through which to represent (or challenge) the nation and forge a collective identity (Delany 1992; Kaplan 1994; Sandberg 2003; Tangherlini, this volume). To extend this focus even further, the real spaces of memory (cf. Jacobs 1996) or symbolic places increasingly include cemeteries, monuments, and memorials, often themselves symbolically located in places where actual conflicts or events took place (Kellerman and Azaryahu 1999). Hung, for example, describes the way Tiananmen Square—itself a site of significant symbolic political meaning—has become "a prime visual means of political rhetoric in modern China," thus creating a veritable "war of monuments in the Square" (1991, 85).

These critical approaches to geography rely on an understanding of spaces and places through the images, symbols, and representations that give them social, cultural, and political meaning, which in turn galvanizes and transcends their existence as purely spatial entities. As Shields (1991) summarizes, these new critical geographies are those in which myths and images add a socially constructed level of meaning to places and themselves help constitute and reconstitute places as meaningful to different people. The idea that socially and culturally meaningful spaces and places—as well as politically contentious places—are as significant as and interact in complex ways with physical spaces is central to the chapters in this volume.

Korea's Critical Geographies

The challenge to exploration of critical geographies is particularly pertinent in Asia, where transnationalism and globalization are being played out with profound spatial and cultural consequences. One such consequence is the marked increase in the movement of people throughout the region, bringing groups from disparate cultures into closer proximity, either in the real world or in the virtual world. Strongly related to this increasing mobility of people, cultures, and symbols, globalization has also provided a catalyst for the reemergence of concerns of cultural homogenization, national identity, and neocolonialism in East Asia. Many of the responses to these changes are being expressed variously through social movements and democratic struggles, ethnic/national and cultural revivals, and the forging of new, hybrid cultures. These changes all occur in space, and many are fundamentally about space itself. New uses and meanings are attached to places both physically (as they are occupied and reconstructed) and figuratively (as they are reimagined, reinterpreted, and articulated). These geographies emerge to challenge, subvert, and resist other dominant meanings in ways that have far-reaching consequences. Consequently, the way that people interpret the space around them and enact new geographies is a critical issue in our understanding of the current changes in Asia. Although these processes inform the interaction of peoples throughout the region, we propose here an intensive study of these processes in Korea, both from a synchronic and a diachronic perspective. Developments in Korea allow us to focus explicitly on some of these areas in which critical geographies are being expressed. As communities experience rapid change with the influx of populations from other regions or countries, the potential interpretations of newly shared spaces multiply and, in some cases, become charged with conflict. In some cases, the appropriation and subversion of sites are much more sustained and seemingly passive, and they do not derive from transnational movements of people. Rather the movements of ideas across boundaries, often through newly emergent media such as those enabled by ubiquitous and inexpensive access to the Internet, also have the potential to radically refigure the landscape. Punks in Seoul, for example, take over an underground café, converting the space into their own "nation," despite the intended uses of that same space by the building's owners (Epstein 2000; Epstein and Tangherlini 2002). Even when places seem clearly defined and their uses without question, one often finds moments when the space is redefined, such as the Tiananmen uprising in China, in which a square dedicated to the symbolic representation of state authority suddenly became the site for a direct challenge to that state.

Korea is no stranger to such confrontation and challenges to authoritarian definitions of space. For example, the well-known massacre in Kwangju mapped onto the city the state apparatus of control, while the opponents of brutal state oppression briefly refigured the landscape of authority in their usurpation of key buildings in the city. Other challenges to state authority that were intimately linked to a geography of resistance took place in Korea throughout the 1980s. Myŏngdong Cathedral—located in central Seoul near one of the most fashionable shopping districts in the city—became a center and refuge for antigovernment protesters, so that violent clashes with police, in which the cathedral was a backdrop, frequently spilled into the upscale pedestrian streets of the area. Myŏngdong Cathedral has consequently become a central symbol of democratization in the country. During the 1970s and 1980s, the cathedral harbored political dissidents, including labor and student movement leaders. It has also been the location of countless prodemocratic protests, rallies, and hunger strikes. Although it has not yet been subject to official reinscription and politicization by the Korean state, it has become a popular tourist site and continues to act as a "terrain of resistance" (cf. Routledge 1993) through the staging of migrant worker protests. Indeed, the division of Korea itself inscribes onto the peninsula a cartography of control that violently alters historic patterns of circulation, land use, and notions of region.

Yet even these spaces of resistance and symbolic sites of opposition and defiance cannot be read in singular terms. Mangwol-dong Cemetery, on the outskirts of Kwangju City, has been the major site in the memorialization of the Kwangju Uprising and a symbolic space for renewing a spirit of resistance to government authority until recently. Since the completion of the New Mangwol-dong Cemetery in 1997 as part of a process of state-sponsored memorialization of the Kwangju Uprising, the old cemetery, known now as the May 18 Movement Cemetery, has become increasingly obscured and dwarfed by the elaborate concrete testaments to the uprising in the grounds of the new cemetery, officially titled the 5.18 Memorial Cemetery. The appropriation of the cemetery as a memorial site by the Korean state in the mid-1990s has drawn the Kwangju Uprising into the domain of official, sanctioned versions of the uprising. In this process, a narrative of the uprising that sees the event as central in forging the nation's hard-fought struggle for democracy is rapidly becoming canonised. This official reinterpretation of the event has provided a challenge to the meanings and ownership of the event by local people in Kwangju City and South Chŏlla Province themselves (Yea 2002).

Sites of memory, in short, convey political messages—consciously or subconsciously, intended or unintended—which, as Richter remarks, have been only very rarely studied (Richter 1999, 109; Yea 2002, 1555). At other times, challenges to state authority (and attempts by the state to counter these challenges) are more protracted and less violent. The "folklore village" in South Korea, for example, is open to widely divergent readings by the Korean government on the one hand and a disaffected populace on the other; the park simultaneously represents an ideal vision of the nation's past and a history of oppression and inequality (Tangherlini, this volume). This historical cartography is repeated throughout Asia in similar manners and is not specific to Korea; yet each example is embedded not only in the historical and political exigencies of its national context.

Tourist sites and festivals geared toward a presentation of a master narrative of "Korea and Korean culture" to outsiders also enable these competing claims to space and at times contradictory understandings of place. Similarly, the recently constructed Comfort Women Museum in Kwangju, Kyunggi Province, allows a hidden history of sexual violence and suffering to be exposed and consumed. The museum is devoted to an exploration and representation of the history of former military sex slaves, or comfort women, serving Japanese soldiers immediately prior to and throughout the Second World War. The main purpose of the museum, according to those who run it, is to expose this previously hidden history through visual and other sensory displays, and to educate Japanese and Korean citizens about this deep wound in the history of these two nations. Unlike the folklore village and other state-authored national memorial sites, the Comfort Women Museum does not constitute a state-authored text aimed at constructing a popularly consumed national identity. Rather, it seeks to interrogate such constructions by publicly acknowledging an undercurrent of national shame and suffering (Yea 2003).

While national identity has been the subject of much academic attention in Korea recently, regional and localized identities have been relatively neglected. Yet regionalism and economic development are also frequently inscribed onto politicized landscapes; a critical reading of the landscape opens the possibility for an understanding of the contours of geographies of resistance that are part of the larger critical project presented here. Land use issues are intimately linked to notions of identity, so when farmers confront developers over contentious issues that pit continued agricultural output against housing development, it is not simply a matter of competing economic interests. These struggles are highly emotive precisely because the meaning of the land differs enormously between groups and because the land is, in each case, bound up in the dialectic tension between projected notions of "progress," cultural identity, and the abilities of the poor to stave off the pressures of the wealthy. These are but examples of the types of tensions and diverse loci in which "critical geographies" move into the foreground.

Organization of the Volume

This volume brings together scholars from disparate fields of cultural studies, geography, ethnography, anthropology, folklore, history, and literature in exploring the meanings/readings, contestations/subversions, and authorship/ownership of space and place in Korea. The essays in this volume are arranged topically, but this is a deliberately informal arrangement.

The perspectives of the authors and the topics they have chosen to address emerge in dialogue with one another. At times, the authors elucidate and amplify certain concepts that resonate throughout the volume, and at other times they contradict and contest theoretical premises and proposed readings of complex spaces. The arrangement of the essays is not intended to be proscriptive, and certainly they could be read in any order. Rather, the idea behind the current organization is to emphasize some of the key concepts that emerge through reading a series of essays together. The rubrics are intended to highlight some of the central commonalities between small groupings of essays, as well as to reveal the inadequacy of any such exclusionary typologies.The section "Geographies of the (Colonial) City" focuses on Seoul during the Japanese Colonial occupation from 1910 to 1945 and the lasting impact of that period on the construction of specific places in Seoul. While other essays refer to this formative period in both the design of urban space and the mapping of the cultural space of the peninsula, the two essays grouped here focus specifically on this time period.

The two essays in the section "Geographies of the (Imagined) Village" delve into the implications of recent economic and industrial development on the conceptions of the village, both in constructed, representational space such as the Korean Folk Village and in rural villages that were physically transformed through the processes of rapid modernization.

Religion and religious practices have always informed the interpretation of Korean geography. The principles of *p'ungsu,* for example, inform not only urban planning but also the siting of religious places. The essays in the section "Geographies of Religion" reveal how religious sites are both historically and environmentally contested and how sites themselves have a great degree of mobility. The politics of religious practice in Korea and the mobility of the sites of that practice emerge in the far-ranging essays in this section.

Places that exist at the margins are powerful loci for the negotiation of identity and aspects of cultural ideology. The section "Geographies of the Margin" focuses on places that exist at the margins of Korean society. While the Seoul Train Station sits squarely in the center of Seoul, it was transformed during the international monetary fund (IMF) crisis into a site where the homeless congregated. Although invoking the notion of "village," the *kijich'on* (or camptown villages, found near US military bases) are excised from mainstream circulation and exist as a locus where foreign contact and trafficking define the landscape. The virtual space of the Internet—cyberspace—has also emerged as a marginal realm where normative aspects of culture are contested and refigured. That refiguration echoes in the constructed urban landscape as well.

In the opening essay, "Respatializing Chosŏn's Royal Capital," Todd Henry shows how urban planners refigured the spatial dimensions of Seoul and the implications the resulting urban landscape had, not only on the circulation of people, but also—and perhaps more importantly—on the use and understanding of open, public space. The implications of the Japanese colonial era and Japanese scholars' fascination with Korean ancient history resonates through Jong-Heon Jin's essay on Kyongbok Palace and the Government General building. Using the demolition of the former Government General building in 1996 as a starting point, Jin analyzes the multiple, conflicting discourses of nation and history that informed the decision to tear down the building. Built on the site of the Chosŏn dynasty central palace, Kyongbokkung, the Government General building was used for various purposes, eventually housing the National Museum before meeting its ultimate fate. Jin details how various groups lobbied to either preserve, demolish, or postpone demolishing the building and how these contrasting positions on a question of the man-made environment resonated throughout the country.

The discussion of the National Museum acts as a bridge to the next section. Timothy Tangherlini in his essay explores another museum space, that of the Korean Folk Village (Minsokch'on). Although the Folk Village was constructed to present the Korean rural past in a specific manner—a manner that aligned with government policies of rapid industrialization—interpretation of the constructed village space was not easily contained. Later interpretations of the village by visitors eventually included a radical departure from the intended narrative, rereading the environment as a geography of resistance. The *minjung* (the people) reading of the Korean rural landscape reemerges in David J. Nemeth's critical exploration of the transformation of rural villages in Korea. The "New Village Movement," launched in the early 1970s by Park Chung-hee's government, radically transformed the physical construction of Korean villages, without significant consideration for the people living in those villages. Later *minjung* art from the 1980s revealed a deep dissatisfaction among certain groups with the new rural landscape that elided aspects of identity, suggesting that the modernization of the villages was an act of environmental violence as extreme as the driving of iron bars into the ground by the Japanese colonial powers to disrupt the circulation of ki (vital energy) (Jin and Kendall this volume). In each of these cases, the projected, imagined concept of the village did not align with the constantly shifting terrain of the dynamic Korean countryside.

The connection between people and the earth also informs religious interpretations of the environment to a profoud degree in Korea. Religious sites in Korea are often overlapping, deeply contested, and reveal the extraordinarily rich history of religious practice in Korea. An example of such a contested site is Kyeryong Mountain. Je-Hun Ryu provides a thorough investigation of the various uses, including military, that have dramatically affected how people "use" the mountain as a sacred site. Mountains play an important role in the imaginary repertoire of practitioners of many of Korea's religions—in particular, shamanism and Buddhism. Laurel Kendall, in her exploration of modern Korean shaman shrines, traces a marked mobility of sacred sites and mountain shrines that are refigured, replaced, removed, and reconstructed by and around government policies, the goals of developers, and the religious ideas of the practitioners themselves. Her intriguing case study reveals how shamans and their customers reconcile themselves and their sacred landscapes with other at times conflicting demands of ongoing development. The importance of mountains as sites of religious practice emerges in Robert Oppenheim's exploration of Kyŏngju Namsan as well. Echoing ideas that emerge in the essays in the section on "Geographies of the (Colonial) City," Oppenheim traces how Kyŏngju Namsan has emerged as a central environmental trope in the understanding not only of Buddhist practice but also in the understanding of local history. The efforts of individual scholars, local and national societies and organizations, and the practices of visitors to the mountain all conspire to create a rich, deeply textured, yet often contradictory narrative of place in contemporary Korea.

In the final section, the city and the relationship between the city and the "outside"—be it a geographic outside, such as the city, or a conceptual outside, such as the categories of "foreigner," "sexual orientation," or even "social class"—emerges as an important geographic concept. In her essay, Jesook Song reveals the transformation of the plaza in front of the Seoul Train Station during the economic crisis (1998–2001) popularly labeled in Korea as the IMF crisis. The government policies toward the homeless—and the classification of the homeless into different categories—influenced to a great degree the spatial practices of the homeless. Congregating at the front of the Seoul Train Station, a building erected during the Japanese colonial period, the newly homeless brought to the very center of the nation's capital a stark reminder of the implications of Korea's economic policies and forced what would otherwise have been kept at the margins to the utter core of Korean life.

Keeping things at the margins, both physically and conceptually, also informs the policies surrounding the US Army bases and the *kijich'on* that have grown up near those bases. Basing her essay on intensive fieldwork in the *kijich'on,* Sallie Yea offers a rare perspective into the economies of the "villages" and the government policies that support these landscapes of foreign incursion, transgression, sexual exploitation and, paradoxically, of hope (however faint) and possibility. Another landscape of hope and possibility emerges in cyberspace, where an otherwise marginalized segment of the Korean population, lesbian-gay-bisexual-transgendered, has been able to create a sense of community. Michael Pettid in his essay shows how this virtual community has rendered itself physical with the construction of bars and other meeting places in Seoul specifically aimed at this population. All of these essays also speak to new forms of mobility—class, transnational, and gender—that have had significant interaction with the physical and cultural environments.

In short, the current volume attempts to investigate aspects of Korean cultural geography from a series of interrelated critical perspectives. While there exists an increasing body of work that focuses on "geographies of resistance," as well as the tactical use (or interpretation) of space as a means for disenfranchised populations to counteract the goals of a dominant group, the majority of these studies focus on issues or events in Europe and Africa (Pile 1997). As scholars of East Asia in general—and Korea in particular—it is our responsibility to interrogate the critical endeavor that proposes an easy mapping of analytic methods and theoretical conjectures derived from a study of European or American phenomena onto East Asia. We believe that in an era of globalization, failing to pay attention to the specific contours of these phenomena might lull us into a sense that there truly has been a leveling of cultural processes, essentially making us as scholars complicit in a postcolonial theoretical gesture that effectively erases the local. Even those studies that do include articles on the Pacific Rim generally include only a single case study or two from East Asia and rarely include mention of Korea. This approach necessarily obscures the fact that—given the high degree of circulation of populations throughout Asia and the intimate economic and historical ties between the countries of the area—considerations of space and the interpretation of place are closely linked throughout the region. Other studies that focus on the rapid economic and political changes within Asia tend to be constrained by adherence to a political economy perspective, muting questions of culture and space (Dirlik 1993). This volume is the first of what we hope will be many to systematically address the complex issues surrounding space, its use, and its interpretation within this clearly defined geographic and political region.

Part 1

Geographies of the (Colonial) City

2

Respatializing Chosŏn's Royal Capital

The Politics of Japanese Urban Reforms in Early Colonial Seoul, 1905–1919

Todd A. Henry

The Urban Reconstruction of a "Living Past"

Among his many brilliant observations, Henri Lefebvre once posited the simple but illuminating argument that "schematically speaking, each society offers up its own peculiar space, as it were, as an 'object' for analysis and overall theoretical explication" (1991, 31). As a French Marxist sociologist writing in the wake of the student-led protests of 1968, Lefebvre aimed his critical energy at theorizing the spatial forms taken by post-Fordist capitalism. In doing so, he sought to highlight the revolutionary potential of urban inhabitants to remake city space within the realm of what he called everyday life.[1] Although he focused on analyzing an explicitly capitalist space, Lefebvre's above-mentioned observation suggests new ways of considering space as a specific sociocultural product requiring critical examination. In what follows, I use the terms "view from above" and "view from below" as heuristic devices for talking about two different yet interrelated spatial scales. The view from above, on the one hand, includes what Lefebvre calls *spatial practices* (the process of producing the material form of social spatiality as perceived by measurement and description) and *representations of space* (the conceptualized space of scientists, planners, urbanists, technocratic subdividers, and social engineers that is tied to the relations of production and, especially, to the order or design that they impose). The view from below, on the other hand, encompasses the view from above, but instead it foregrounds what Lefebvre has termed *spaces of representation* (space as directly lived through its associations, images, and symbols and hence the space of inhabitants and users) (18–19).

This essay draws on the insights of Lefebvre in order to analyze the creation and re-creation of Japanese colonial space in Seoul, which had served as the royal seat of the Chosŏn dynasty (1392–1910) before becoming the capital city of Japan's empire in colonial Korea (1910–1945).[2] My discussion here is limited to the early colonial period—that is, from the onset of the protectorate in 1905 to the end of the first decade of colonial rule in 1919. In setting this temporal frame, which deliberately straddles the 1910 divide, I want to suggest the importance of exploring how Japanese officials actively remade the spaces of the royal capital even as they selectively destroyed them. As I will suggest, *Japanese* efforts by the Government-General to respatialize Seoul into the colonial city of Keijō (in Korean [K]: Kyŏngsŏng) intersected with earlier *Korean* attempts by the Great Han Empire (1897–1910) to refashion Chosŏn's royal city of Hanyang into the imperial capital of Hwangsŏng, in an effort to create an autonomous, monarch-centered nation-state.[3] In critically analyzing how Japanese urban planners sought to modify the city's preexisting spatial configuration "from above," I will also show how these attempted respatializations involved an active engagement with rather than simple dismissal of the lives of the Korean residents and Japanese settlers. Furthermore, these inhabitants did not respond passively to the impositions of the new colonial state; rather, they actively sought to protect their individual and collective interests, not all of which the Government-General could effectively redirect. As I will demonstrate through a discussion of these contestations over city space "from below," their interests tended to collide with those of the colonial government in the reconstruction of what I will call "officially sanctioned social" sites, which included Keijō's main thoroughfares as well as its parks and other civic sites.

The tensions endemic to these historical engagements with space reflect a central contradiction that existed between the two strands of colonial discourse and practice defining the respatialization of early colonial Seoul and, more generally, Japanese colonialism during this period (1905–1919). On the one hand, I identify a "civilizational" strand, based on a universalized temporal gauge of progress. In the context of my discussion, this strand involved the attempt to mark *radical difference* between colonizing Japanese and their Korean others.[4] This process of Japanese colonial othering will become most evident in the discourse on Koreans' lack of "civic morality" (in Japanese [J]: *kōtokushin;* K: *kongdŏksim*), a charge frequently leveled in the face of noncooperation with urban reforms. This discourse sought to juxtapose such "uncivilized" Korean behavior against the "proper" civic morality of Japanese. In practice, however, the hegemony inherent in such representations was repeatedly disrupted by complaints about selfish Japanese settlers who evaded their civic obligations and withheld their property from the new colonial state. Government leaders also lamented the tendency of Japanese settlers to disparage and mistreat the Korean residents of colonial Seoul, whom official ideology regarded as subjects of a "benevolent" Japanese emperor, to be treated as equals.

This policing of the ambivalent position of the Japanese settler community is closely tied to the other strand of colonial discourse and practice, which aimed to incorporate the city and its inhabitants into the greater empire. Within this incorporative strand, I identify two substrands. One consisted of the somewhat ambiguous policy of what I call Japanese style "assimilation" (J: *dōka;* K: *tonghwa*). Often invoked under the sign *isshi dōjin* (K: *ilsi tongin*), this incorporative ideal "conveyed the [Confucian-inspired] idea that all who came under the sway of the [Japanese] sovereign shared equally in his benevolence" (Peattie 1984, 97). The prominence of this substrand expanded dramatically following the explosive anticolonial uprising of March 1919, and it is best illustrated by the adoption of a more clearly defined policy of assimilation by the governor-general, Saitō Makoto (1919–1927 and 1929–1931). In its many forms, assimilation thus aimed to transform Koreans into dutiful subjects of an imperial community centered on the Japanese emperor. The other substrand, meanwhile, was distinguished by a more universal pattern of globalizing colonial capital; it involved constructing a planned network of widened and sanitized roads that could clear the way for the efficient circulation of goods and people. Here in particular, Japanese colonial discourse and practice originating in the "civilizational" strand was rapidly forced to cope with the everyday exigencies of administering the new colonial city. As a result, this administrative project did not mark radical difference between Japanese and Koreans and thus exclude the colonized from the time and space of Japanese colonial modernity; rather, it was based on the incorporation of colonized Koreans as hard-working and dutiful imperial subjects.

This deep ambiguity took concrete shape in the profoundly uneven spaces of early colonial Seoul. To be sure, the first decade of Japanese rule did see the emergence of a partially upgraded grid pattern of widened, straightened, and sanitized roads. However, despite painstaking efforts to control and direct Koreans, the colonial officials of this decade achieved only limited success in transforming the microspaces of Korean city life, whose rhythms did not always synchronize with the new logic of Japanese colonial rule. In most cases, grudging acquiescence, creative adaptation, or active negotiation bridged the gaps opened by the pressures of everyday life in the new city. But when these measures no longer functioned, Koreans took to the streets to directly challenge the spaces in and through which Japanese colonial rule operated, as the concluding discussion of the March 1919 uprising illustrates.

From Royal Hanyang to Imperial Hwangsŏng: A Brief Look at Precolonial Seoul

To date, Korean postliberation/nationalist descriptions of mid-to-late Chosŏn period Seoul have overwhelmingly focused on the implications of the city's socioeconomic development for the intellectual and practical purposes of establishing an autonomous nation-state. The historiographical component of this project emphasized the significant modernization of late Chosŏn Seoul, only to have the roots of its "internal development" (K: *naejejŏk palchŏn*) extirpated by Japanese colonialism (Eckert 1996). Authors of this persuasion aimed at countering Japanese theories of stagnation, which had by contrast downplayed the economic potential of the precolonial city. These colonialist accounts instead focused on the politico-symbolic vacuity of Hanyang/Hwangsŏng's major architectural monuments—the city wall, its gates, and palaces—and the pressing need to transform them (cf. Shinbu Junpei 1901, 47, 117). Without accepting all of the highly polemical implications of either Japanese colonialist theories of stagnation or Korean postliberation/nationalist theories of internal development, one can nonetheless identify significant changes in the role played by late Chosŏn Hanyang—most notably that the capital city came to function *both* as a politico-symbolic center and as a commercial hub.

According to Ko Tong-hwan, both the national circulation of metallic currency and the implementation of the Uniform Land Tax Law during the latter half of the seventeenth century spurred the development of an urban economy based on commercial currency (1997, 98–100). Poor harvests and epidemics during this period drew many disparate peasants to the expanding suburban areas just outside the city's walls, thereby increasing the metropolitan population from just over 80,000 in 1657 to nearly 200,000 in 1669, a level it retained until the end of the dynasty (103). By the late eighteenth century, the extramural population of merchants and other social groups nearly matched that of the city's official residents living within the city's walls. Within the city's walls, the commoner markets of Ihyŏn and Ch'ilp'ae came to complement the luxury and everyday products needed by the royal palaces and sold at the Six Licensed Stores (K: Yugŭijŏn) located around Chongno. Accordingly, the majority of Hanyang's population was engaged in some form of commercial enterprise by the eighteenth century, and new, nonelite forms of culture and entertainment developed around the middling classes of the so-called *chungin*—bureaucratic specialists in foreign languages, law, and medicine (Hwang 2004; Kang Myŏng-gwan 1992; Ko 1999, 327–372).

With the onset of imperialist aggression during the late nineteenth century, protecting the economic and political autonomy of the dynasty became a growing concern for Chosŏn's political elite. Following the Sino-

Japanese rivalry over the peninsula, the assassination of Queen Min by the Japanese military, and King Kojong's flight to the Russian Legation in 1896, concerned officials of the newly established Great Han Empire worked to re-create the royal city of Hanyang into an "imperial capital" (Hwangsŏng) as part of a larger project to create an autonomous, modern nation-state. This campaign included the following projects (Fig. 2.1): (1) the destruction of temporary commercial stalls (K: *kaga*) that jutted from the city's main thoroughfares, thereby restoring their original width; (2) the creation of new roads centered around the Taehan Gate of the Kyŏng'un (later

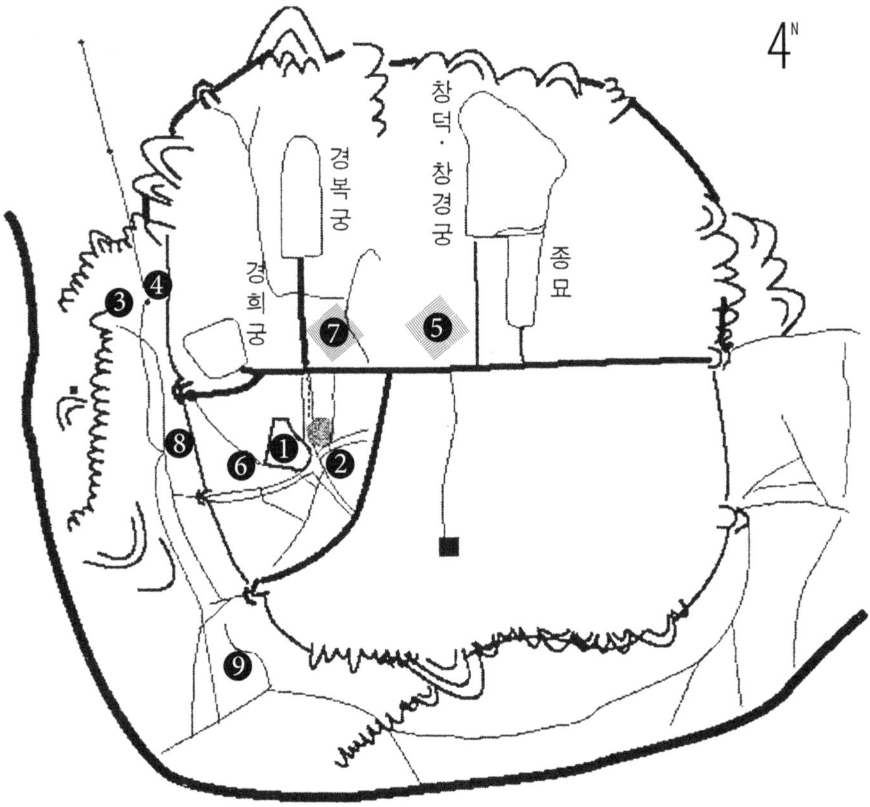

Fig. 2.1. Urban reform projects during the Great Han Empire (1897–1910):

(1) Kyŏng'un (Tŏksu) Palace
(2) The Wŏn'gu Altar
(3) Independence Gate
(4) Independence Park
(5) Pagoda Park
(6) Public Park
(7) Memorial Monument
(8) Keijō (Train) Station
(9) Yongsan

Tŏksu) Palace (#1 in Fig. 2.1) that connected to preexisting roads, thereby establishing a radial system of roads that linked the new imperial capital to the city's outskirts; and (3) along the city's roads, the construction of buildings and structures of political importance, such as the Independence Gate (#3), Pagoda Park (#5), and a Memorial Monument (#7) (Yi 1999, 97–98).

Recent Korean scholarship has shown that this plan for the construction of imperial Hwangsŏng likely adopted Washington, DC, as its conceptual model. Despite the tacit acknowledgment of this external stimulus, a strong historiographical predisposition toward internal development has prevented Korean scholars from fully acknowledging the city's embeddedness in the semicolonial situation of the Great Han Empire (1897–1910) and protectorate (1905–1910) periods, not to mention a discussion of the long-term consequences of this imperial/colonial situation. Yi T'ae-jin, a proponent of the Washington, DC, model thesis, has gone so far as to argue that "Japanese colonial rule mercilessly cut off the full realization of the ingenious plan that harmonized both the old and the new in establishing a modern and independent Great Han Empire" (1999, 120).

By contrast, a more transnational understanding of turn-of-the-century Seoul shows that the leadership of the Great Han Empire was engaged in a highly international process of nation-state building, the "native" and "non-native" elements of which are not easily disaggregated. On the one hand, some Korean elites sought to establish greater national autonomy by downplaying the influence of China on the country's past and thereby promoting Korean cultural primacy. Andre Schmid has described these nationalizing initiatives as the "de-centering" of the Middle Kingdom (2002, 55–100). The symbolic valence of this project was concretely manifested in the construction of the Independence Gate, overwriting the site on which Chinese envoys had been regularly received by the Chosŏn court (Yŏngŭn Gate). It was also reflected in the erection of the Hwan'gu Altar, built for new, national ceremonies on the very site of those envoys' temporary residence in Hanyang (Nambyŏng Palace). While these projects to create an imperial capital exuded a new self-confidence in state nationalism, they were carried out with and against the imperial powers of the post–Sino-Japanese War, particularly Russia and Japan.

Indeed, the very space of Hwangsŏng came to reflect the precarious geopolitical position in which the Great Han Empire found itself. It was not by coincidence that the center of the new imperial capital, Kyŏng'un (later Tŏksu) Palace, was constructed in close proximity to the foreign legations. In his pioneering work on the Great Han Empire's urban planning projects, Kim Kwang-u, in fact, uncovered that a secret passageway and a special bridge were built to link the Russian Legation (where Kojong briefly resided after the Sino-Japanese War) to the new imperial palace complex

(1990, 115).[5] Other modernizing urban projects were likewise tied to the semicolonial politics of concessions during this period. For example, the Great Han Empire employed two American entrepreneurs—Henry Collbran and H. R. Bostwick—to introduce new technology and to finance the construction of a streetcar, power lines, street lamps, water pipes, and telephone lines (No 1980). Like the internal passageways of the imperial palace complex, the streetcar line was also symbolically laid from Kyŏng'un Palace, across the main thoroughfare, Chongno, to Ch'ŏngnyangni on the eastern outskirts of the city.

Apart from the international implications of Hwangsŏng's perilous position within the politics of East Asian imperialism, the transnational character of this predicament also ushered in a new domestic relationship between the elite officials who managed a modernizing state and the nonelite members of the Korean nation. In the context of the city, the construction of an imperial capital at Hwangsŏng formed part of a government-led program to create a symbolic national center from and through which to integrate the previously socially stratified groups of Koreans into national subjects of Emperor Kojong. In this regard, it is noteworthy that the streetcar line mentioned above linked the imperial palace complex with the commercial district along Chongno and the newly established civic site at Pagoda Park. As Chŏn U-yong has convincingly argued, the period of the Great Han Empire marked a break with the past, when virtually the entire city consisted of "royal spaces." By contrast, the new imperial city of Hwangsŏng witnessed a noticeable increase in the number of contact zones between the sovereign and the people (1999b, 40–41).[6] These zones included the area in front of the Taehan Gate (the entrance to Kyŏng'un Palace), where the city's residents increasingly gathered for new national events, such as Kojong's 1897 elevation from king to emperor. In spite of their Confucian rhetoric calling for a "popularly oriented" (K: *minbon;* J: *minpon*) nation, the kind of Korean "people" envisioned by the elite architects of the Great Han Empire were, in fact, closer to "subjects" (K: *sinmin;* J: *shinmin*)—charged with duties to strengthen the state in the face of national crisis—rather than free and autonomous "citizens" (K: *simin;* J: *shimin*) endowed with individual rights. When some Korean elites associated with the Independence Club pushed for a more participatory constitutional system, the leaders at the Korean court quickly moved to disband this organization and to centralize state authority through an emperor-centered system of national government (Chandra 1988).

With Japan's victory against Russia in 1905 and the subsequent establishment of a semicolonial protectorate government, the monarch-centered project to transform the city into a national center became nearly impossible, especially after Kojong's forced abdication in 1907. However, popular sentiments toward the throne remained alive in the minds of the city's Korean residents, for reasons to be discussed in the conclusion. And although the Korean-led attempts to reconfigure the city were ultimately thwarted by the Japanese annexation of the peninsula, the *modern* project of physically and symbolically transforming the capital, as initiated by the leaders of the Great Han Empire, remained an ongoing task even after 1910. The major change following the annexation was, of course, that thereafter a foreign-dominated colonial state undertook the difficult and coercive task of incorporating the city's Korean residents as subjects of the Japanese emperor.

From Hwangsŏng to Keijō: Constructing a Japanese Colonial Capital

With the promulgation of the Annexation Treaty on August 29, 1910, Japanese authorities moved to gain full control over the symbolic topography not just of Seoul, but of the entire peninsula. To this end, they changed the name of the colony back to Chōsen from its previous designation of Taehan Cheguk, or the Great Han Empire, a name associated with a semiautonomous and nationalizing peninsula under the rule of Emperor Kojong. In addition to Taehan, the use of the name Hwangsŏng, which designated the former imperial capital of the Great Han Empire, was also prohibited (Chong 1980, 62, 70). The city was symbolically renamed Keijō, invoking the character for "capital," as had been the case when Edo became Tokyo in 1868 and replaced Kyoto as the new Japanese imperial capital. Like Tokyo—which did not emerge as the nation's symbolic center until 1889 (due to the historical weight of Kyoto and the emperor's early Meiji peregrinations outside of the new capital)—the process of transforming the royal/imperial capital of Hanyang/Hwangsŏng into the colonial capital of Keijō was neither immediate nor uncontested (Fujitani 1996, 31–92). Although seeking to erase the traces of the former Great Han Empire, Japanese colonial policy makers were forced to engage with its ongoing relevance in the colonial present. This reliance on the living Korean past was, in fact, irrevocably written into the very language of the Imperial Rescript announcing the annexation: "Even after annexation, his majesty, the Korean Emperor [J: *Kankoku kōtei heika*], along with nobles below him shall be treated with considerable favor. The well-being of the people, standing directly under his [the Japanese Emperor's] tutelage [J: *suifu*], shall improve, and industry and trade will show marked progress, [now] that they are under control" (Imperial rescript 1910). On the one hand, this reference to the Korean royal house was a reflection of the limitations of early Japanese imperial power to reach the Korean people, who, particularly in Seoul, had been increasingly exposed to the nationalized language, symbols, and signs of the Great Han Empire. On the other hand, the invocation of the Japanese emperor bespoke the ambitious goal of eventually assimilating Koreans as colonized subjects.

This project began with the strategic reconstruction of the city's palaces, whose grounds as private royal spaces officials converted into civic parks and other public monuments. Made accessible to the Korean masses for the first time, these symbolic sites were used to draw newly colonized Koreans into an expanding imperial community dominated by colonizing Japanese. Before the annexation, Japanese commentators on the capital city commonly referred to Kyŏng'un (later Tŏksu) Palace, the residence of Kojong, as "the capital of the kingdom" (J: *ōjō;* K: *wangsŏng*) or "the capital of the empire" (J: *kōjō;* K: *hwangsŏng*), and Kyŏngbok Palace as the "old capital" (J: *kyū-ōjō;* K: *ku-wangsŏng*) (Ōta 2003, 207).[7] Protectorate-era efforts to move this "imperial capital" elsewhere bore fruit when, in 1907, the puppet emperor Sunjong was forced to live in Ch'angdŏk Palace, soon transformed into a colonial park. To further deflect attention away from the Korean imperial center, Japanese officials quickly moved to destroy or sell the buildings related to Kyŏnghŭi Palace, which, along with Kyŏng'un Palace, made up the ruling palace complex of the Great Han Empire.[8] Parts of Kyŏng'un Palace—symbolically renamed Tŏksu Palace after Kojong's ouster in 1907—remained, but it was stripped of its emperor and the powerful symbolism he wielded. Officials built an art museum on the palace grounds, which they subsequently surrounded by a complex of Japanese colonial structures.[9] A similar fate awaited Kyŏngbok Palace, the central palace of the former Chosŏn dynasty. As part of a nearly fifteen-year plan to build the new Government-General Building (completed in 1926), the formerly closed ceremonial grounds of Kyŏngbok Palace were converted into a functional and periodically open site, where in 1915, for example, officials convened the Korean Industrial Exhibition (J: Chōsen bussan kyōshinkai; K: Chosŏn mulsan kongjinhoe) in order to commemorate the first five years of colonial rule.[10]

In addition to the selective de-/reconstruction of Seoul's royal/imperial palace grounds, the early colonial regime also sought to rearrange the city's roads and neighborhoods. These and other urban reforms undertaken during this period were referred to by the Japanese expression *shiku kaisei,* which literally means "reforming the city's districts." First carried out in early Meiji Tokyo, *shiku kaisei* connoted a rudimentary form of urban planning: widening and straightening extant roads, expanding waterways for sewerage, and refashioning domainal and religious spaces into civic parks (J: *kōen*) and plazas (J: *hiroba*). It aimed at a partial upgrade of the preexisting city rather than a full, systematized transformation of the city's spaces (Fujimori 1990, 90). As in Meiji Tokyo, one of the initial urban reform projects in early colonial Seoul involved the sanitizing, widening, and straightening of preexisting roads, over which would then be laid a grid system of streets connected by a series of rotaries converging on the Kōganemachi Plaza in the Japanese settlement of southern Seoul (Fig. 2.2). Such an ambitious urban planning project, although considerably scaled back in 1919, had proven impossible in the new Japanese capital, Tokyo, where entrenched landowners prevented its implementation. With fewer such restraints and the aid of a resolute military police force, Japanese officials first implemented this ideal city plan in early colonial Taipei (Gotō 1998, 103–116; Ishida 1987, 51–106; Sorensen 2002, 60–84).

In practice, however, this ambitious urban planning project was modified according to the dual nature of the colonial government's "incorporative" project to assert political, economic, and cultural control over both the

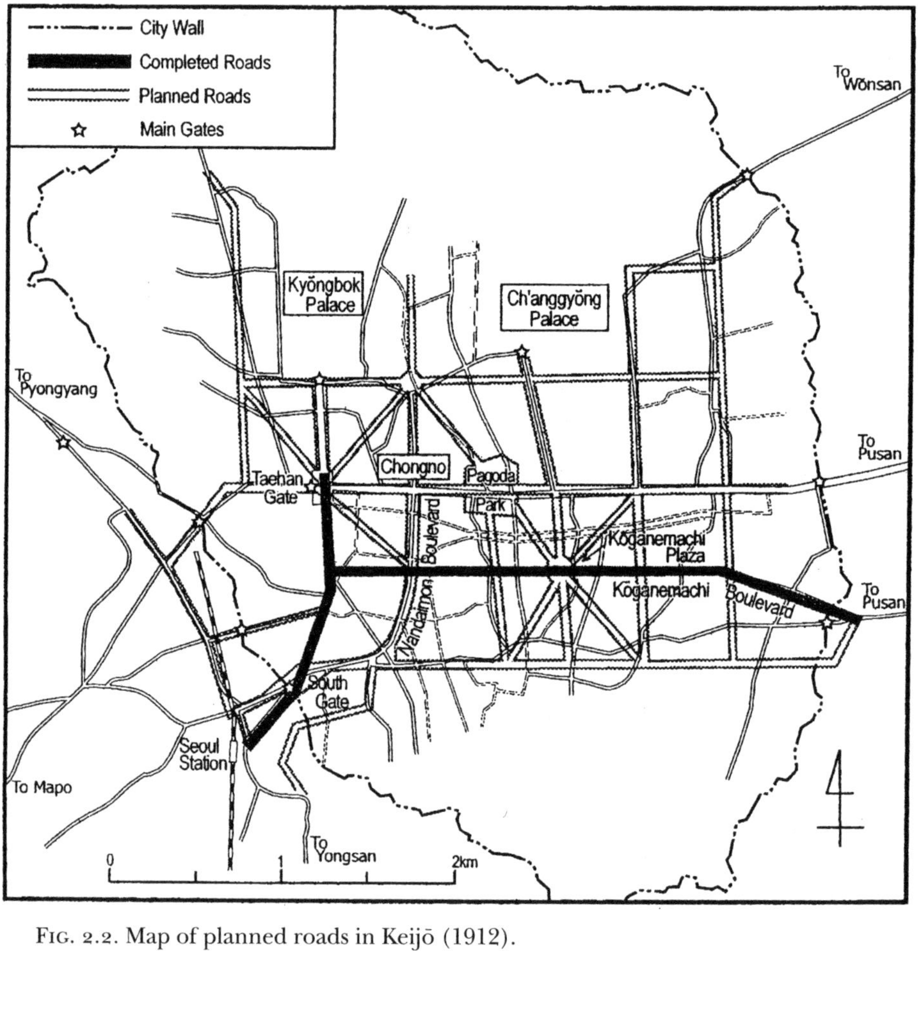

FIG. 2.2. Map of planned roads in Keijō (1912).

Korean neighborhoods (located primarily north of the Ch'ŏnggye stream) and the Japanese settler community (located at the foot of Namsan). To this end, the Government-General, following nearly four years of investigations, promulgated a unified urban administration system in 1914. Two years later, it instituted a system to place both Koreans and Japanese under the supervision of officially appointed neighborhood administrators. The 1914 law disbanded the administrative organization of the Japanese settler community, in addition to those of other foreign settlements with extraterritorial privileges in early colonial Korea. Despite its aim at greater administrative efficiency, this move at urban consolidation provoked a backlash by the Japanese settler elite, who desperately sought to maintain a voice in the future of the city. Responding to rumors in 1911 that the Government-General Building would move to Chongno in six or seven years, Fuchigami Tadasuke, director of the settler community's road-widening committee, pleaded with the colonial government to expand Honmachi Avenue (the narrow commercial road running through the Japanese business district) in order to prevent the Japanese center of the city from moving northward (Fuchigami 1911). Although the colonial government rejected Fuchigami's proposal, most (thirty-one) of the roads marked for widening and straightening between 1913 and 1917 were undertaken in the southern half of the city, where the majority of Japanese settlers resided. However, because the colonial government also sought to gain control over the Korean population living in the northern part of the city, one of Keijō's major focal points eventually came to center on the Government-General Building. This symbolic shift in the city's center of gravity from Kōgane-machi Plaza (as manifested in the 1912 plan; see Fig. 2.2) to the soon-to-be-completed Government-General Building is shown in Figure 2.3. Through this spatial reconfiguration, the predominantly Korean-inhabited northern half of the city (K: Pukch'on) was thus connected to the overwhelmingly Japanese-populated southern half of the city (K: Namch'on) along the main axis of Taihei Avenue (today's Sejongno). Following the construction of the long-awaited Government-General Building and Chōsen Shrine, this north-south axis came to reflect Japanese imperial power in its various manifestations: military (Yongsan), sacerdotal (Namsan), economic (Honmachi District), and political (Kyŏngbok Palace grounds).

As this sketch of the urban planning of early colonial Seoul suggests, these attempts to respatialize the city were closely linked to the ambiguously defined policy of Japanese style assimilation. This project, not unlike the Great Han Empire's attempt to unify the Korean people under the auspices of Kojong, involved the difficult task of incorporating these same people as subjects of the Japanese emperor. As one official newspaper article boldly pronounced, invoking the oft-cited, ideological rhetoric of "imperial benevolence":

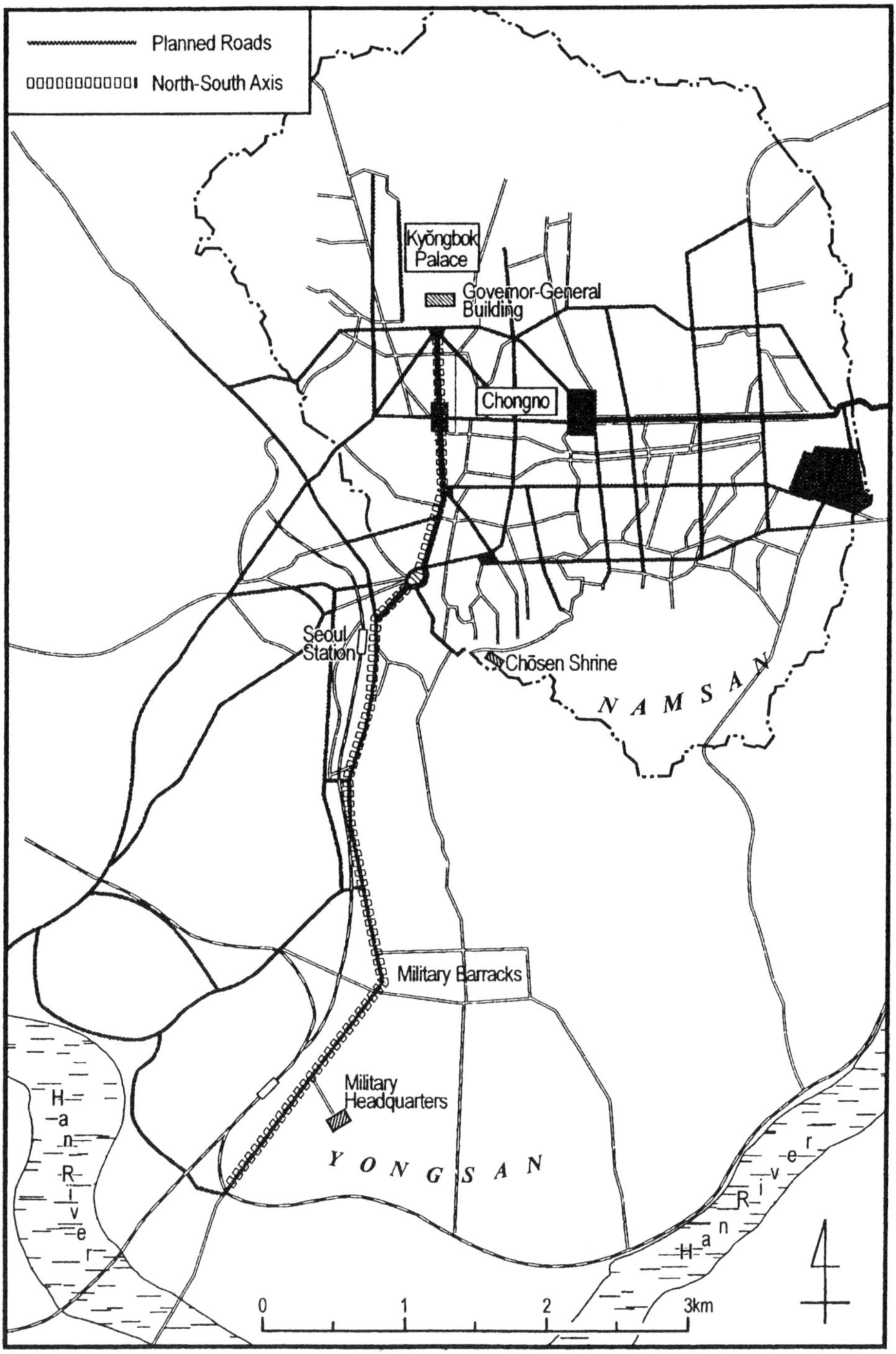

FIG. 2.3. Revised map of planned roads in Keijō (1919).

> Insofar as [both] Japanese *[naejiin]* and Koreans, from the dumb to the smart and from the bright to the benighted, are now equally under the benevolent gaze of the Emperor, rights *[kwŏlli]* shall naturally spread [among them].... With equal *[kŭnil]* knowledge, they shall work together, the myriad duties *[bŏnbaek ŭimu]* [of Koreans] the same as those of the Japanese, enjoying, as a matter of course, the graciously bestowed privileges *[kwŏlli]* of the Emperor (*Maeil sinbo,* "Kyŏngsŏng puminhoe haesan," 1911).

Despite these lofty promises of equal rights, officials invariably stressed the duties and responsibilities of Koreans as subjects of the Japanese emperor. As one Japanese official admitted in terms of the potential limitations of "Japanizing" Koreans, "even if [Koreans] cannot become 'faithful imperial subjects' [J: *teikoku no chūryō naru shimin*], I believe that through moral guidance *[kyōka],* they can become 'dutiful subjects' *[junryō naru shinmin]* of the Empire" (Watanabe and Abe 1988–1991, 23).

The other incorporative strand aimed at channeling the socioeconomic resources of the peninsula into the development of the greater empire. In early colonial Seoul, this project involved constructing a planned network of wide, sanitized roads that would clear the way for the efficient circulation of goods and people. As one newspaper article explained to Koreans about the position of roads in the modern (colonial) city, "Generally speaking, roads have a direct connection with civilized transportation *[munmyŏng kyot'ong]; . . .* if transit is convenient, import and export frequent, and the coming and going of people trouble-free, there will naturally be an unlimited profit *[mujung ŭi iik]*" (*Maeil sinbo,* "Sigu kaejŏng," 1912). However, news reports frequently documented a number of obstacles blocking the potentially "enlightened" use of roads and other city spaces. One 1912 article from the *Maeil sinbo,* the Korean-language newspaper overseen by the Government-General, lamented that Keijō, despite its importance as Chōsen's flourishing capital, could not even boast one straight road. It described the streets as overlapping like snakes, narrow and uneven, making it difficult for people to pass. Even the main thoroughfares, it argued, remained congested by the unregulated erection of temporary stalls (K: *kaga*) (*Maeil sinbo,* "Sigu kaejŏng," 1912).[11] By superimposing a grid system of straightened, widened, and clear streets over these preexisting roads, colonial officials hoped to create a well-organized and prosperous urban infrastructure in which Koreans, along with their Japanese counterparts, could efficiently carry out their duties as diligent colonial subjects, while making a profit at the same time. With the advent of urban electrification, newspaper pundits also suggested that illuminated Korean night markets be established as a way of controlling and taxing itinerant hawkers while at the same time offering them the possibility of increasing their profits.[12]

The modern grid system of roads was also crucial to the related urban project of redrawing the city's administrative districts. Following the pre-

Meiji Japanese tradition of constructing commoner areas in a regular grid pattern of rectangular blocks *(chō)*, government officials attempted to create similar administrative units in colonial Seoul. Figure 2.4 shows this ideal city plan, in which approximately three subblocks *(chōme)* are bisected by a straight road, thus forming one block-shaped administrative unit.

As will be discussed in the next section, however, contestations over the use of city space on the part of colonized Koreans and Japanese settlers, not to mention financial restrictions, forestalled the radical spatial transformation of early colonial Seoul. As a result, only fifteen of the forty-two planned road projects were undertaken, leaving a large part of the precolonial system of roads in place, especially in the back alleys of Chongno and other neighborhoods in the northern half of the city (*Chongnoguji* 1994, 57).[13] This spatial holdover, in turn, ensured that in spite of the Japanese urban ideal of uniform, road-based neighborhoods, the preexisting structure of "natural" neighborhoods prevailed. An example of such neighborhoods—communal living spaces consisting of a number of plots (K: *p'ilji*) divided by a few cul-de-sacs or by one minor (but not necessarily straight) road—is shown in Figure 2.5.[14] In the end, most of the city's Korean neigh-

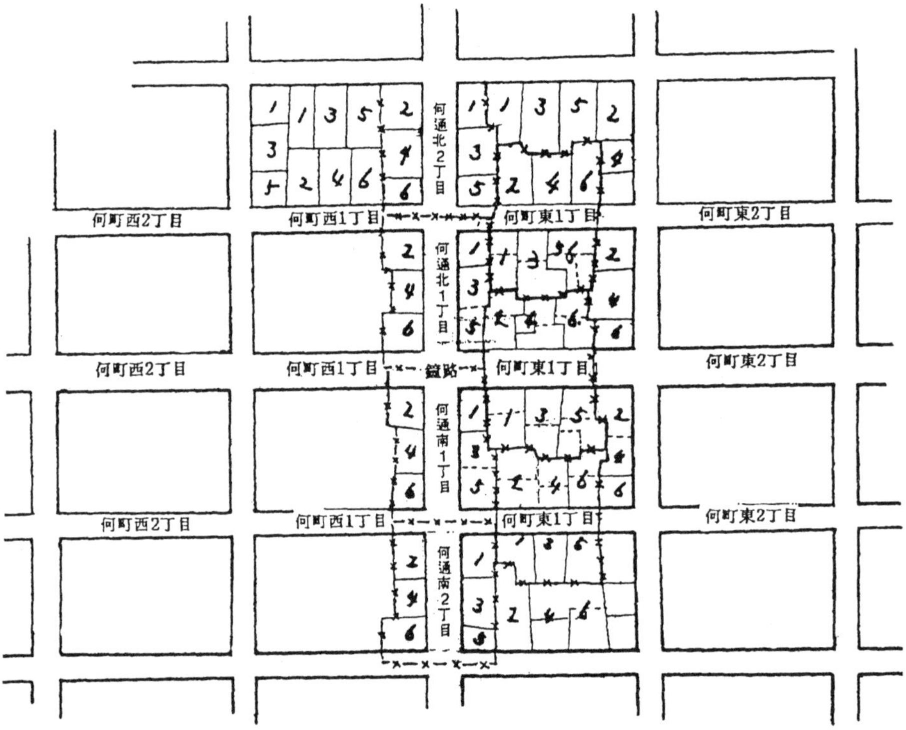

FIG. 2.4. Japanese road-based neighborhood formation.

borhoods retained their precolonial administrative boundaries, even as a loose grid network of thoroughfares took shape around them. As a result, a characteristically uneven spatial structure came to define the early colonial city.

The Politics of Urban Reforms: Contestations over Officially Sanctioned Social Spaces

In order to rationalize Keijō's system of roads, early colonial officials realized that it would be necessary to persuade, if not coerce, Korean and Japanese city dwellers to relinquish their land and property. To this end, the Government-General passed the 1911 Land Confiscation Law (J: Tochi shuyōrei) and a series of other ordinances such as Regulations Controlling Urban Construction (J: Shigaichi kenchiku torishimari kisoku, 1913) and Regulations Controlling Roads (J: Dōro torishimari kisoku, 1913). These laws imposed heavy fines on those who disobeyed a complicated set of regulations pertaining to activities on or near what are best called "officially sanctioned social roads" (J: *kōkyō dōro;* K: *konggong toro*) and other similar places such as civic parks (J: *kōen;* K: *kongwŏn*).[15] In order to avoid

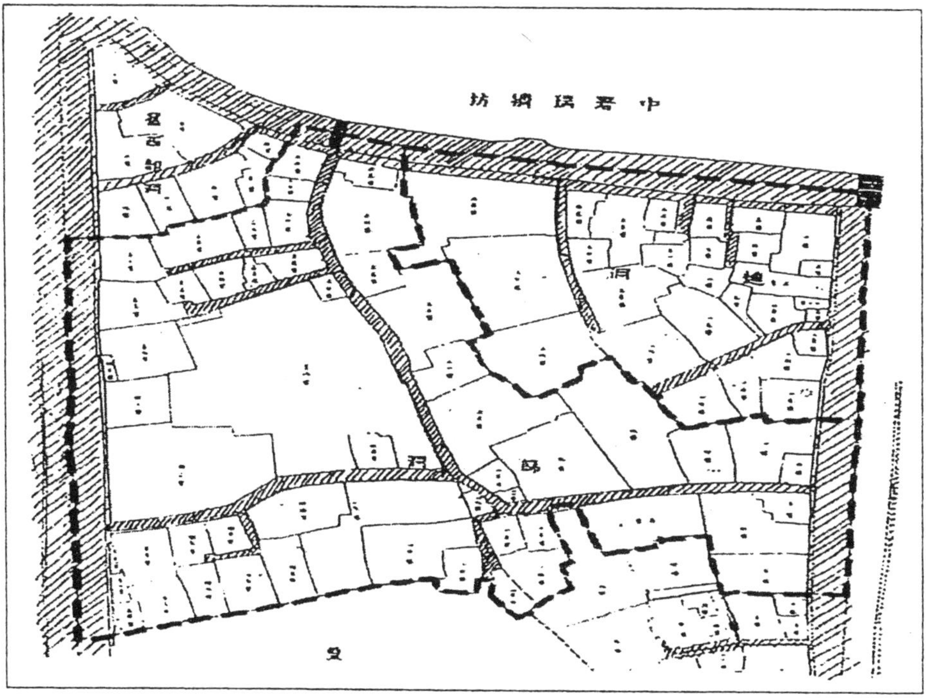

FIG. 2.5. Korean, "natural" neighborhood formation.

the semantic connotations that the word "public" usually conveys today of open access, participatory parity, and social equality, I am using the unwieldy but historicized expression "officially sanctioned social" spaces. With this terminology, I recognize that "private" (J: *watakushi;* K: *sa*) interests were allowed a contingent amount of leeway. At the same time, I concur with Harry Harootunian's suggestion that such interests were "a necessary adjunct" of the "official" (J: *ōyake;* K: *kong*) realm, which dictated that one's "inner self corresponded to the required outer behavior" of Japanese colonial policy (1974, 120, 117).[16] The related term for "public" (J: *kōkyō;* K: *konggong*) captures the ideological dimension of "official" ideology by obfuscating the specter of "private" (selfish) interests with the insertion of an idealized sense of community, conveyed by the character for "togetherness." In practice, *kōkyō* or *konggong* discourse was thus deployed to control the "private" practices of Koreans (and Japanese), which did not necessarily conform to the "official" standards of propriety as expected in specific city spaces. In doing so, Japanese colonial authorities could choose from a range of tactics by which to incorporate (assimilate) the colonized into a hierarchical imperial community of obedient and dutiful subjects: selectively reinforcing certain "desirable" older traditions (filial piety, social hierarchy, and communal aid), discouraging other "undesirable" ones (early marriage, overreliance on others, superstitions, etc.), and encouraging new practices (loyalty and patriotism, diligence and industry, and the efficient use of time and money).[17]

In spite of these efforts, Korean-language newspaper reports from the 1910s repeatedly chastised colonized subjects for not cooperating with the land survey, a project that aimed to establish a rationalized system of property ownership and that was closely tied to constructing a grid system of roads.[18] As one article put it in ideologically dichotomous terms, "Civilized people *[inmin toen cha]* wholeheartedly welcome and approve this [project], but those lacking knowledge *[molchisik]* rely on barbarian traditions *[yamae sŭpgwan]*. . . . Thus, misunderstandings *[ohae]* regarding the land and building surveys are many, as are wild rumors *[puŏn]* that taxes will be increased or that dim-witted people's homes and land will be confiscated without warning" (*Maeil sinbo,* "Kaok t'oji ŭi chŭngmyŏng," 1912). While Japanese officials continued to mock such popular notions as groundless, they were at the same time forced to explain that land could in fact be appropriated by the colonial state for its urban reform projects as long as the owners were properly compensated (*Maeil sinbo,* "Sigu kaejŏng kwa ohae," 1912). Despite these financial inducements, newspaper articles from the 1910s continued to carry admonitory accounts of "stubborn" Koreans. These stalwarts, in contrast to their "civilized" counterparts who surrendered themselves to the "officially sanctioned social good" (K: *kongik*), were accused of overstating the value of their property out of utter self-interest; thus, the articles stated, they "forced" officials to implement the Land Confiscation Law. Japanese urban planners of early colonial Seoul were also frustrated by frequent and unpredictable land transfers, including cases of falsified property sales and even "unfilial" instances wherein Korean sons were said to have sold their fathers' land for their own profit (*Maeil sinbo,* "Kaok t'oji ŭi chŭngmyŏng," 1912).

These accusations point to the colonialist supposition that Koreans, individualistic in nature, lacked any "sense of civic morality" (K: *kongdŏksim;* J: *kōtokushin*). Despite the charge of unfiliality mentioned above, some Japanese observers remarked that Koreans did in fact possess morality (J: *dōtoku;* K: *todŏk*), but one that was excessively centered on the family (or, in the other extreme, on the individual). As Okita Kinjō, a popular Japanese colonial ethnographer, argued about the supposed particularity of Koreans:

> They do not have the slightest feelings for people outside their family. Even if there is a person fallen on the roadside or a neighbor with a nasty illness, Koreans only shed blood and tears within the narrow confines of their families. It is as if there were no need or desire to have an effect on those outside the family. On this point, they are clannish people to the end, and provide no adequate credentials *[tekiseru shikaku]* for being a member of society or the nation *[shakai kokka]* (1905, 50).[19]

According to Okita and other Japanese colonial commentators, this family-based morality had by its very nature prevented the establishment of community facilities such as civic parks, city orphanages, and public libraries. With the gradual construction of such facilities after the annexation, Koreans were urged to use these institutions to cultivate a sense of civic morality that befitted them as modern Japanese imperial subjects. Such exhortations included keeping one's house and the area in front of it sanitary, using officially sanctioned communal facilities with a "spirit of civic propriety" (K: *konggongsim;* J: *kōkyōshin*), and cooperating with other officially led efforts to reform the city's spaces and its residents' attitudes toward them.

Despite these typically colonialist accusations of "lack" and the accompanying exhortations for "improvement," official Japanese discourse on Koreans' attitudes toward the city and its officially sanctioned social spaces was continually strained by reports that Japanese settlers themselves failed to live up to these civic expectations. Such accounts were evident even in official Korean-language reports, which ostensibly aimed at presenting a pristine image of model Japanese residents to the colonized. In an article on the widening of Namdaemun Avenue, for example, the *Maeil sinbo* reported in late 1911 that two Koreans and one Japanese living along this thoroughfare had refused to surrender their land. In contrast to Pak Sŏnho and Yun Ch'ang-sŏk, Koreans who begrudgingly acquiesced when the colonial government threatened to use the Land Confiscation Law, a Japanese man named Hamaoka Ryōtetsu refused to comply, forcing officials to apply legal sanctions. This report, furthermore, complained that Japanese settlers, despite considerable knowledge and wealth, could conduct themselves without common sense (*Maeil sinbo*, "Nanmunoe t'oji suyong," 1911). Compromising accounts of uncooperative Japanese settlers continued to appear in the Korean-language press. An editorial from 1913, for example, unabashedly reported that in contrast to Koreans—fewer and fewer of whom refused to sell their land or did so only at exorbitantly high prices—settlers from the metropole continually committed acts deemed inconsiderate and selfish by the colonial government. How, the author of this article wondered, could Japanese as "forerunners of civilization" *(munmyŏng ŭi sŏnsin)* hold such "perverse ideas" *(pyŏngi han sasang)?* (*Maeil sinbo*, "Sigu kaejŏng kwa konggongsim," 1913).

To gain control over the Japanese settler community and thereby create a greater "unity of interests" between the government and nongovernment sectors, the colonial administration of Terauchi Masatake (1852–1919), the governor-general from 1910 until 1916, resorted in large part to coercive legal and police measures. In 1912, for example, the Government-General passed the harsh Regulations for Criminal Punishment by the Police (J: Keisatsu hanshobatsu kisoku), which revised a similar but less comprehensive law passed in Korea in 1908.[20] Drawing on the language of a similar law instituted in the metropole just a month earlier, the colonial version of this law regulated the "public" behavior of both Japanese settlers and Korean subjects, thereby conforming to officialdom's ideal of Japanese assimilation through "imperial benevolence." According to Lee Chong-min, a scholar of penal reforms in colonial Korea, the Regulations for Criminal Punishment by the Police gave colonial authorities a comparably high degree of latitude in enforcing measures deemed necessary to ensure proper "public safety" (J: *kōan no hoji*), predictable commercial transactions, and a generally sanitary urban environment (2004, 338–347).

In addition to purely coercive measures, pundits also called for more persuasive sociospatial means in order to mobilize Japanese settlers in the colonial project of assimilating the city's Korean residents. Matsui Shigeru (1866–1945), a police bureaucrat serving in the Japanese protectorate government in Korea, called for the establishment in Seoul of what he called "social education facilities" (J: *shakaiteki kyōiku kikan*): parks, theaters, music halls, clubs, libraries, art museums, zoological gardens, and so on. These civil institutions, he contended, were the most appropriate way to "harmonize public feelings *[jinshin o yūwa]*, to bring about taste *[shumi]* among the people, and, by guiding them with sophistication *[yūga ni michibiki]*, to foster the establishment of civic morality *[kōtoku]* and thus to bring about harmonious co-operative life *[enman naru kyōdō seikatsu]*" (Matsui 1913, 830). With regard to the residents of early colonial Seoul, he argued with no lack of imperial Confucian paternalism that such institutions should form part of a larger, two-step assimilatory effort: first, to "civilize" the Japanese residents of the city who, as "elder brothers" *(anibun),* would then "lead and direct" *(yūdō shiji)* their younger Korean siblings toward "civilization" *(kaimei)* with "kindness and care" *(shinsetsu to teinei)* (846). After annexation, this debate over how to improve the "taste" of the city continued unabatedly, with calls for the establishment of a high-brow theater *(koshō naru engekijō),* a proper club *(kanzen naru kurabu),* and a major Shinto shrine.[21] Although the Japanese male elite residents' motivations in pushing for these facilities likely diverged from those of their counterparts in the Government-General, they disclose, nevertheless, a shared class- and gender-specific sensibility of what constitutes social and cultural activities "proper" Japanese should pursue. In practice, however, this idealized sensibility did not necessarily correspond with the reality of life in the early colonial city, where low-class Japanese peddlers roamed the streets for a profit and a relatively high percentage of women from the main islands found themselves outside of the home in the booming entertainment districts.[22] Thus, despite official calls for nonelite settlers to join in the "civilizing mission" mentioned above, in practice the colonial police and military were left to enforce "proper standards" of civic morality among both unruly Japanese and incompliant Koreans.

This was particularly the case in urban sites that, according to official policy, were reconstructed to serve both recreational and educational purposes. Take, for example, the place of Pagoda Park. In contrast to the highly symbolic colonial park strategically built on the grounds of Ch'angdŏk Palace, Pagoda Park was the only major civic park located in the predominantly Korean neighborhoods of Chongno. Completed around 1900 on the grounds of Wŏngak Temple, this park, like other urban reform projects carried out during the turn of the century, was constructed under the auspices of the Great Han Empire (1897–1910) but aided by foreign advisors. Accessible only on Mondays during the first three years of colonial rule, Pagoda Park was finally opened to the public on a daily basis on July 28, 1913, having been upgraded with benches, flowers, walking paths, electrical lighting, and trees (Kang Sin-yong 2004, 56). Despite the addition of what colonial officials considered the necessary accoutrements of a modern (Japanese) park, newspaper articles from the period repeatedly criticized Koreans' allegedly improper use of this place. These allegations included the uprooting of trees, the damaging of foliage, and so on. Ignoring the material privations that underlay these "uncivilized" acts, colonial officials assailed such behavior, seeking to control it by exhorting Koreans to cultivate a sprit of civic morality. In doing so, they paid particular attention to what they referred to as the "private" nature of Pagoda Park during the precolonial period—namely, its limited hours of operation and restrictions on visitors' usage. These limitations were diametrically juxtaposed against the self-proclaimed openness of the reconstructed "modern" park, although this apparent freedom was circumscribed by various provisions of using a Japanese colonial park. As one official Korean-language newspaper article explained to the colonized:

> Generally speaking, parks, regardless of the conditions of their construction, are not the private property of an individual *[ilkaein ŭi samul];* rather, they belong to the people of the "officially sanctioned collective" *[konggong].* Hence, they should be carefully tended to with civic spirit *[konggong ŭi maum].* If . . . this spirit of tender care *[aeho ŭi maum]* is weak or damaged, it is fair to say this person is without civic morality *[kongdŏksim]* (*Maeil sinbo,* 1914).

Although repeatedly dismissive of behavior deemed improper at such sites, newspaper articles such as this one almost always failed to specify how this vague notion of "civic spirit" was to be positively cultivated. In addition to discouraging "improper" uses of parks, colonial officials also urged Koreans to remove commercial and residential structures that encroached on the city's officially sanctioned social roads and build new ones elsewhere. The itinerant merchants (K: *kaga sangin*) of the South Gate area had been partially incorporated into the city's nearby central market, which was established by the Great Han Empire in 1897 to block Japan's commercial penetration of this area of the city. Plans were also made during this period to remove the temporary stalls along Chongno and install a new streetcar, although the fate of this area's merchants remains unclear. Judging from the dramatic drop in Seoul's 1910 population, it is possible that some (if not most) of them left the city for the increasingly impoverished suburbs or for the countryside (Chŏn U-yong 1999). Despite both the financial and emotional dislocations that would accompany such selfless acts, the bottom line in all of these exhortations was, of course, to positively contribute to the construction of a harmonious, Japanese-led, imperial community.

Exhortatory on the one hand, this civic discourse—epitomized by the modern repackaging of the Confucian slogan to "give priority to 'official' concerns over 'private' ones"—also disclosed an obvious gap between the ideal of incorporating the city's Korean residents into a hierarchical imperial community and the practical exigencies of early colonial life amidst intense pressures to obey a myriad of new and likely confusing regulations (Lee Chong-min 2001, 95–130). In practice, the cultivation of this organic connection between civic-minded Koreans and the greater imperial community suffered from an inherent tension: Most colonized Koreans, although burdened with duties and responsibilities like Japanese settlers, lacked the political and economic rights (also unequally) bestowed upon ethnic Japanese citizens. In order to conceal this apparent contradiction in colonial policy, officials resorted to exhortations encouraging Koreans to first and foremost work hard to carry out their duties and responsibilities as imperial subjects. According to one newspaper editorial with the lofty title "The sanctity of labor," these travails would be rewarded *at some unspecified, later date* with treatment and rights bestowed upon "first-class citizens" *(iltŭng kungmin)* (*Maeil sinbo,* 1914). As Leo Ching has noted, it was this "continuous *deferment*" in the materialization of the Japanese project of emperor-centered assimilation that functioned as "the ideology par excellence for concealing the gap between political and economic discrimination and cultural assimilation" (2001, 113, 106).

By Way of Conclusion: Pagoda Park, March 1919

Japanese colonial ideology might have functioned *discursively* to conceal this gap; however, *in practice,* it was always in danger of being exposed by the actions of colonized Koreans, most of whom from the start lacked the material and emotional resources that they were asked to invest in the creation of an imperial community. Although the examples mentioned above—refusing to sell one's property needed to build a new road or uprooting a tree from a park to use as firewood—can be considered creative adaptations to the pressures of everyday life in the early colonial city, at certain moments urban practice moved beyond such individual tactics and came to involve more violent collective actions. By way of conclusion, I would like to return to Pagoda Park and the infamous example of March 1919 to illustrate this point. As discussed above, this park formed part of a series of officially sanctioned social sites, created not just for recreation but also to encourage the city's residents to cultivate a "proper" sense of Japanese civic morality. Temporally, 1919 coincided with the close of the first decade of formal colonial rule, historiographically categorized as one of "military rule" (J: *budan seiji;* K: *mudan chŏngch'i*). In spite or because of the draconian nature of the Government-General's efforts during the 1910s to forge a compliant and "civic-minded" colonial urban populace, Pagoda Park came to play a central role in the massive peninsula-wide uprising of March 1, 1919 (Baldwin 1979, 123–162). Although unable to attend to all the delicate historical and historiographical complexities of this uprising here, I shall briefly discuss this site/moment for what it illuminates about the early colonial history of space and place making discussed above.

In his study on the practice of everyday life, Michel de Certeau argues that the "tactic," as he calls it, "belongs to the other," and that it "insinuates itself into the other's place, fragmentarily, without taking it over entirely, without being able to keep it at a distance." He continues, "Because it does not have a place, a tactic depends on time—it is always on the watch for opportunities that must be seized 'on the wing'" (1981, xix).[23] In the context of the 1919 uprising, this "place" that most Koreans lacked was a viable institution through which to express their needs and wants. Seoul's palaces, which might have served as sites for national expression as conceived of during the Great Han Empire, had been strategically transformed into colonial parks in an effort to cultivate new loyalties to the Japanese Empire. In spite of these efforts, popular sentiments toward the Korean throne remained alive after 1910, partly due to the fact that early colonial policy makers continued to rely on the former royal house as a symbolic example of the Government-General's intention to assimilate the colonized. As part of this policy, the colonial government resurrected the protectorate-period custom of royal processions wherein Sunjong, the former and last Korean emperor, traveled along the major thoroughfares of Seoul to visit his father, Kojong, at Tŏksu (formerly Kyŏng'un) Palace.[24]

It was no coincidence, then, that the March 1st uprising coincided with the death of the leader of the former Great Han Empire, Kojong, whose corpse rallied thousands of Koreans to demand alternatives to the martial style of Japanese rule they had endured over the previous decade. Reappropriating Pagoda Park, the leaders of the movement transformed this officially sanctioned social place as a counter site to the incorporative logic of Japanese colonialism, using it instead to declare their independence. Throngs of protestors then proceeded in an equally significant act to march along the main thoroughfares of the colonial city, the very roads the Government-General had recently reconstructed to facilitate colonial rule. As they dispersed to sites of importance throughout Keijō, one large group stopped at the symbolic center of the former Great Han Empire, the Taehan Gate of Tŏksu Palace, where they yelled *"Manse!"* or "Long live the king!" one last time.[25] Through these symbolic gestures, the participants of the March First Movement seized upon and exploited the contradictions inherent in the early colonial respatialization of Chosŏn's royal capital, particularly the myth of equal treatment under a "benevolent" emperor. Although only temporarily transforming those urban spaces through which the Government-General sought to control the colonized, the Korean protestors succeeded in questioning the harsh terms of their incorporation into the imperial community, forcing officials to rethink their strategy of colonial assimilation.

Notes

Research for this essay was made possible by a grant for field research from the Korea Foundation. I would like to thank Ch'oe Pyŏng-t'aek and Chŏng Sang-u, both graduate students in modern Korean history at Seoul National University, for helping me to read and make sense of the *Maeil sinbo;* any errors in interpretation are my own. John Duncan, Namhee Lee, and Micah Auerback offered constructive criticisms of my argument and its development.

1. For more on the oeuvre of Henri Lefebvre, see Merrifield (2006).

2. The Korean term "Chosŏn," the name indicating the period of Yi dynasty rule, will be distinguished from the Japanese word "Chōsen," which was used to designate the status of colonial Korea (1910–1945). For convenience, I will also use the English expression "colonial Korea" to refer to Chōsen.

3. On the nature of these reforms, see Schmid (2002, 72–80).

4. Although an argument for "ancestral sameness" (J: *nis-sen dōsoron;* K: *il-sŏn tongjoron*) was used as an argument for the annexation of the peninsula and occasionally as an ongoing justification for colonial rule, one cannot detect as central to 1910s Japanese colonial discourse or practice what Oguma Eiji has called the "ambivalent sameness" (rather than radical difference), which would later come to characterize Japanese style assimilation (J: *dōka;* K: *tonghwa*). As the theory advocating the Japanese as a "mixed ethnos" (J: *kongō minzoku*) suggested, the colonized eventually came to be situated in a "blood relationship" (J: *ketsuen kankei*) based on the Japanese "adopted child" (J: *yōshi*) system, which rationalized and sustained the emperor system by incorporating colonial subjects into the imperial family (Oguma 1995). On the Japanese colonial government's policy forbidding Koreans to use Japanese names prior to 1930, see Mizuno (2002). On the discourse of ancestral sameness during the early colonial period, see Chŏng Sang-u (2001, 183–231).

5. On the history of this district (Chŏngdong), see Kim Chŏng-dong (2004).

6. On the conceptual foundations of this phenomenon in the era before Kojong, see Yi T'ae-jin (2000, 11–46).

7. That Japanese commentators used the character for "castle" and not "palace" when referring to Seoul's palaces shows, according to Ōta, that they approached these spaces as the early Meiji government had the former *daimyō*'s castle complexes, which were also converted into parks, schools, and other "public" spaces (2003).

8. It was here where Korean officials, under increasing pressure from the Japanese-led protectorate government, tried, albeit unsuccessfully, to hold a national exposition in 1908 (Ōta 2003, 210–212). A year earlier, the Japanese protectorate government had taken the upper hand in this "exhibitionary competition" by staging the Seoul Exposition (J: Keijō hakurankai) in Tonghyŏn/Meiji-machi, which was located away from the ruling palace complex and closer to the Japanese settlement. On the Seoul exhibition, see Mok (2000, 17–20).

9. For more on the fate of this palace, see Chŏng Chae-jŏng (1997, 261–293).

10. On the colonial transformation of the Kyŏngbok Palace grounds, see Son Chŏng-mok (1996, 520–560), Hŏ (1996), and Kim Chŏng-dong (2000, 183–227).

11. According to another article, the author spuriously argued that historically speaking, Korean roads had been laid according to the ill-positioned buildings rather than vice versa (*Maeil sinbo,* "Sigaji kŏnch'uk ch'uich'e," 1913).

12. See, for example, *Maeil sinbo* ("Yasi ŭi p'ilyo," 1910; "Kyŏngsŏng kaksangjŏm," 1911).

13. The number of roads slated for reconstruction increased to forty-seven in 1930, although only twenty-five of them were ever completed (Chongnoguji 1994, 57). On the fate of Seoul's pre-1920 planning, see Kim Ki-ho (1995, 41–66).

14. Comparing precolonial and early colonial land register maps made by the Great Han Empire and the Government-General, Seo Hyŏn-ju has shown that of the fifty-five "natural" Korean neighborhoods appearing on the former, only eight (or 14.5 percent) were reconstructed in the latter as Japanese-style block neighborhoods (2002, 133). On the fate of Seoul's "natural" neighborhoods during the colonial period, see Yang Sŭng-u (2003, 37–71).

15. For these laws, see *Keimu ihō* no. 10 (1911, 158–162); no. 43 (1913, 897–900); and no. 49 (1913, 967–971).

16. As Ishida Osamu has pointed out, the difficulty of assessing the internalized sense of loyalty *(chūjitsushin)* associated with Japanese-style assimilation necessitated that standards for judging the level of one's Japaneseness be linked to external behavior (1998, 51).

17. On the nature of these reforms, see *Maeil sinbo* ("Minp'ung kaesŏn," 1917).

18. On the land survey in early colonial Seoul, see Kang Pyŏng-sik (1994, 27–130).

19. On early colonial Koreans' alleged lack of "civic morality" as related to hygienic practice, see my discussion of Okita (Henry 2005, 643–653).

20. For the 1912 law, see *Taihi chūkai keisatsu hansho batsurei* (1920, 293–327). For the 1908 law, see *Genkō teikoku keiji hōten* (1909, appendix, 11–19).

21. See, for example, Ika ni seba shumika shiuruka (*Chōsen oyobi manshū,* March 1, 1915, 830).

22. On the place of metropolitan women in the urban centers of the Japanese Empire, see Brooks (2005).

23. Henri Lefebvre, also an analyst of everyday life, made a remarkably similar comment in referring to representational space: "It embraces the loci of passion, of action and of lived situations, and thus immediately implies *time*" (1991, 42) (my italics).

24. Tsukiashi Tatsuhiko has shown that these processions actually increased after annexation (2000, 73–75).

25. For the other groups' activities in Seoul, see Kim Chin-bong (2000, 110–117).

3

Demolishing Colony

The Demolition of the Old Government-General Building of Chosŏn

Jong-Heon Jin

In this essay, I examine the nationwide controversy that occurred around 1993–1995 over the question of whether the historic Government-General Building, formerly the seat of the Japanese colonial government in South Korea (1910–1945), should be demolished or not.[1] After about three years of heated debate, this legacy of the colonial period was finally demolished in 1996. The empirical focus of this chapter is on the diverse historical references of the demolition project, which was intended to restore the "national spirit" through the mobilization of urban rhetoric. Further, I explore the discursive contestation among diverse social groups over the project and the consequential modification of the official narratives of the nation.

The social debate shows how official rhetoric of symbolic landscapes is articulated through the struggle to construct national identity and how such politically imposed spatial rhetoric is negotiated with the unoffical/popular memory of landscape and place. I explore the redefinition (or transformation) of the hegemonic narrative of the nation—*anticommunist/developmentalist discourse*—which culminated in the postwar military regimes (1961–1987), tracing it back to the iconoclastic Western-oriented nationalists of the elightenment movement in the early twentieth century under Japanese rule. I will show how, in the course of the controversy, the official discourse was challenged by a newly driven nationalist narrative. The discourse of national identity is not a unitary and monolithic structure of fixed narrative. The transforming landscape of power in Seoul reflects the contested and heterogeneous nature of the national narratives of Korea. It works as an active medium through which the discursive field of the nation is restructured.

In recent years, academics have examined or briefly commented on the meaning of the controversy and the building's ultimate destruction (Bae 2002; Pai 2000; Yoon 2001). Bae provides a postcolonial view of the demolition project, critically examining Kim Young-sam's political rhetoric: the slogans of "rectification of history (Yŏksabaroseugi)," "New Korea (Sinhankuk)," and "Globalization (Segyehwa)." She argues that the destruction of the building is the attempt to create the image of Korea "through the mirror of foreigners" (2002, 264). Her viewpoint eventually tends to reiterate the conventional postcolonial criticism of the "lack of automony" among the colonized. In addition, it seems that some people in academia interpret the building's destruction in the context of their criticism of Korean nationalism (Pai 2000). In differing from these views, I assume that the recent controversy over the colonial building manifests the disruption of the official nationalist discourse and the reformulation of the national identity, rather than the outcry of the existing colonial sentiments or distorted postcolonial reaction.

To examine this, I use some key concepts of cultural geography: the landscape as text and imagery. Yoon (2001) suggests a geographical viewpoint of the building's demolition based on the concept of landscape as text. He examines the general relevance of the thesis of naturalization of landscape from the opening of the Chosŏn dynasty through the modern era. Instead, I focus on a particular historical period—modern times—and elucidate how the urban landscape functions as a communicative device through which specific social discourses and competing nationalist narratives are constituted, reinforced, and contested. First, I will introduce briefly the background of the two historic structures, Kyŏngbok Palace and the Government-General Building, to show how the two landscapes have been symbolically represented in the context of the modern history of Korea. I examine the way in which the Government-General Building was constructed through colonial spatial politics, eradicating the memory of the dynastic past. In the following sections, I discuss the controversy and its discursive and visual contestation.

Historical Background

Kyŏngbok Palace, the symbol of kingship in the Chosŏn dynasty (1392–1909), is located in the center of Seoul, the capital city of Korea. It was originally built in 1392 as the main palace of the Chosŏn dynasty and comprised approximately three hundred traditional buildings, most of which burned down during the Japanese invasion in the sixteenth century (1592–1598). The palace was restored to its original state in the late ninteenth century as part of the reform policy of Taewŏngun.[2] The reconstruction lasted from 1865 through 1868. However, after the annexation of Chosŏn by Japan in 1910, the Japanese colonial government destroyed the majority of the palace to build the Government-General Building at the site. When Japan was defeated in 1945, only thirty-six buildings remained. The destruction of the palace was ordered by Terauchi, the first governor-general of Chosŏn, and implemented as part of the whole spatial transformation project of Seoul, which aimed to position Seoul in the basement of colonial rule (Ryu 2000, 237).

The urban reform was rhetorical and political as well as modern and functional. Based on the Urban Improvement Ordinance promulgated in 1912, the colonial government established a grid system of staightened and widened streets in Seoul, modifying the narrow and meandering flows of the premodern urban network (see Henry 2004 and this volume; Bae 2002, 86–97). In addition, at the northern and southern ends of the newly constructed north-south axis of the spatial structure, two political, symbolic structures were placed: the Government-General Building in front of Kyŏngbok Palace and the Chosŏn shrine at Namsan. This politically choreographed, modern urban rhetoric effectively eliminated the symbolic spatial structure of the Chosŏn dynasty, constructed according to the principle of *p'ungsu*.[3] Thus, the demolition of Kyŏngbok Palace and its gate, Kwanghwamun, became an effective rhetorical device, symbolizing the fatal downfall of the Chosŏn dynasty by utterly denying the inherited symbolic geography, *p'ungsu,* and severing the flow of authority believed to come from the sacred natural landscape.

Considering the history of the Government-General Building, it is ironic that the building remained standing after Korea became independent in 1945, continued to be used as the government office building of South Korea until 1986, and was then transformed into the National Museum until Kim Yŏng-sam's government destroyed it in 1995. The main reason for the reuse of the colonial building, despite its imperialistic implications, was the economic hardship of the newly independent country. The material history of the building is accompanied by several layers of ideological history, which reworked the accumulated social memories of the colonial experience, ideological conflicts, and military dictatorship. During Japanese colonial rule, the building was the key icon of the imperial invasion and dominance. During its use as a government office (Capitol Hall) in the period of the military regimes, it was a symbol of developmentalist and anticommunist ideology, which was supported by South Koreans who shared the agonizing memory of the Korean War (1950–1953) and the confrontational geopolitical division of the peninsula into two countries. Subsequently, the Kim Yŏng-sam administration attempted to visualize a new version of nationhood through the elimination of the remaining legacy of colonialism. The building thus became an unstable embodiment of power relations and popular memory because of its historically differentiated representations.

Demystifying Dynastic Symbols: Making Palaces into Spaces for Public Display

The Government-General Building (Chosŏn Ch'ongdokbu Ch'ŏngsa) was used as the National Museum from 1985 through the time of its destruction. Conservationists in the debate strongly emphasized the fact that the building was the National Museum and no longer the Government-General Building. In fact, the entwined history of landscapes of power and space for display originated from the *politics of display* of the Japanese colonial government, which aimed to demystify the palaces of Chosŏn as royal symbols.

The spatial politics of the colonial regime within the landscape of power were implemented in two ways. They not only restructured the existing spatial arrangement for the construction of the Government-General Building but also transformed royal palaces into modern and functional spaces. These two lines of urban rhetoric were internally connected to each other. Both explain how the "regime of modern vision" was imposed and projected onto the colonized territory, integrating a mode of the panoptic-like surveillance vision and spectators' observing eyes. A towering vertical structure, the colonial headquarters imposed modern, visual authority and power onto the urban space. The transformed royal palaces satisfied people's desire for visual amusement in a "world-as-exhibition" (Mitchell 1989). The royal palace was deprived of its political authority and opened to the public, exposing them to a modern aesthetic vision of the past.

The landscape of dynastic power was thus utterly dislocated, relocated, and transformed into a space for other purposes. Ultimately, the royal palaces were used for public displays such as exhibitions, a zoo, a botanical garden, and an art museum. All of these displaying spaces were the reproduction of a particularly Western way of viewing the world that prevailed in the last decade of the nineteenth century, a "rendering up of the world as a thing to be viewed" (Mitchell 1989, 221).

From September 11 through October 31, 1915, Chosŏn Mulsan Kongjinhoe (the Competitive Products Exhibition) was held in Kyŏngbok Palace in commemoration of the first five years of Japanese rule. This was one year before the construction of the Government-General Building began at the same site. Kongjinhoe was an imitation of other world's fairs that prevailed in major Western cities at the turn of the century. With the appearance of the modern entertainment industry, aided by the advent of photography, trains, and advertising, the world's fair dramatically influenced the change in capitalist societies and public life in urban cultures responding to global issues (Brekenridge 1989, 200–201). These fairs played an important role in the production of symbolic urban landscapes, such as Paris' Eiffel Tower for the Exposition of 1889.

The first international exhibition was launched at the Crystal Palace in

London in 1851. One of the key initiatives of world's fairs in the middle and late nineteenth century was to formulate national culture with institutionalized aesthetics by displaying and viewing rare and uncommon objects from all over the world. The world's fairs were also spaces where the colonizers and the colonized were contrasted and specific forms of colonial knowledge were produced. It is has been pointed out that they were "cultural paradoxes of imperialism" (Brekenridge 1989, 214). Even though the world's fairs stand for "the ecumene beyond national boundaries," they consequently reinforced the image of the non-European world as the "Other."

Terauchi began to prepare Kongjinhoe in 1913 and decided to hold it in Kyŏngbok Palace the following year. Two-thirds of the palace space was allotted for exhibition, and it was divided into eighteen pavilions, including the First Pavilion (agriculture), the Second Pavilion (economy, civil engineering, medical, etc.), and other special pavilions (Hong 2004; Ju 2003). The goal of the exhibition was to produce colonial and modern discourse through the quantified display of progress and development achieved under Japanese rule (Ju 2003).

According to the statistics, Kongjinhoe was quite successful. The number of visitors to the exhibition, including those from Japan, was 1,160,000. This was a huge number, considering that the population of Korea at the time was only 16 million (Ju 2003, 159). The exhibition was a nationwide event. The railways of Honam-sŏn, running from Kwangju (Chŏlla-do) to Seoul, and Kyŏngwŏn-sŏn, running from Wŏnsan (east coast, Hamgyŏng-do) to Seoul and enabling modern tourists to reach Mt. Kamgangsan, were opened for the exhibition. The Government-General gave discounts for train tickets and lodgings in Seoul to exhibitors to encourage the participation of as many people as possible (Hŏ 1996, 69; Hong 2004, 21). Through the nationally constructed railroad system, visitors came from all parts of the peninsula. It was the first mass movement of people for a modern type of collective amusement. The short story *Kongjinhoe,* written by Ahn Kuksŏn and published the year of the exhibition, portrays it as a popular and interesting attraction for ordinary people.[4]

Kongjinhoe served as an overwhelming landscape where Japanese cultural hegemony was spatially expressed and visually/performatively exerted. The exhibition was literally *instructional* as well as amusement oriented. For the first time, visitors were exposed to the modern sense of simultaneous time and space. Tens of thousands of items from foreign countries such as Japan and Taiwan, as well as from all over the peninsula, were competitively displayed. The items showed a sharp contrast between past and present, the ancient regime and the five years of modernized Japanese rule, as well as the earlier colonization of Taiwan. For example, when breeds of rice from Korea and Japan were compared, they found the Japanese breeds superior, and they were recommended for use. The geographical difference was reduced to a unilinear time sensation, which alluded to the possibility of the colonized Chosŏn making such progress through the prepared path to modern society.

Whereas the world's fairs in Western metropolises ultimately showed the contrast between the virtues of civilization and savage life, Kongjinhoe in a sense visualized the linear path of progress and development in a tangible form and in a more instructional way.[5] This difference is related to the contradictory oscillation of Japanese imperialist ideology and political strategies. On the one hand, Japan separated itself from other Asian countries, rendering them as "Japan's orient" (Tanaka 1993). On the other hand, Japan emphasized the shared racial origin of Korea and Japan, which is well articulated in Naisen Itti policy of later periods of the colonial rule. This dual nature of Japanese colonial ideology was represented in a rudimentary form in Kongjinhoe. The exotic view of Kyŏnghoeru (royal garden in Kyŏngbok Palace) and Kisaeng (Keisha) in the advertising poster of the exhibition (see Ju 2003, 150) symbolizes the Japanese version of Orientalism, differentiating traditional Korean culture from Japanese modern civilization. At the same time, Kongjinhoe articulated the possibility of progress and the development of the colonized society rather than irreconcilable differences between the colonizer and the colonized. The old regime (Chosŏn dynasty) was contrasted with the new regime (ten years of Japanese rule) in all aspects of social and economic life. All the changes were presented through statistical evidence, and the dualistic contrast was explicitly drawn in the poster of the exhibition and the structure of the exhibition space (Ju 2003, 150).

Needless to say, spectators in Kongjinhoe were implicitly forced to shed their traditional legacy for the modernized future. In this sense, there could not be a better place for the exhibition than the royal palaces. What was displayed and compared was not only the formal exhibition items but also Kyŏngbok Palace itself. A majority of the palace was destroyed to build pavilions, and the remaining buildings were used for provisional exhibition space, formal ceremonies, and guest rooms for special visitors (Hong 2004, 22). As a result, the surviving remnants of the palace seemed miserable in comparison to the wonders of the exhibition's modern scientific world. The fall of the Chosŏn dynasty was naturally ascribed to its inability to react to the progress of the outer world. With the destruction of the palace and the construction of the Government-General Building, the colonial discourse was visually inscribed on the urban space. Though Terauchi expressed regret for not having built a symbolic monument for the exhibition such as the Eiffel Tower for the Paris Exhibition (Hŏ 1996), he did not recognize that there could be no better monument for the exhibition than the Government-General Building itself. The palace was an effective material and ideological counterpart of the modern/colonial discourse proposed in the exhibition. The palace was not protected, nor was it simply demolished and removed from the collective memory of Korea. Rather, it was systematically and persistently brought into the burgeoning discursive field of nation and modernity, acting as a persuasive antonym for allegedly favorable virtues, such as modern progress and a nation's self-reliance. In this way, the iconoclasm of early enlightenment nationalists was rather exaggeratingly realized, and the exhibition was a symbolic space and time where modern myth of nation as *developmentalism* and its unbeatable virtue of *progress* were being founded as an archetype.[6]

Iconoclasm Revisited, but Reversely Positioned

Iconoclasm was revitalized in the mid-1990s by the first nonmilitary government of Korea since the 1961 coup. The Government-General Building was not the only building demolished by the government's policy drive. Other buildings in Seoul, so-called legacies of the past, were destroyed during this period, including the former residence of the Government-General during the colonial period located in Chŏngwadae (the current president's office) and the apartment complex for foreigners on the slopes of Mt. Namsan. The three destroyed buildings had two elements in common: They served as reminders of foreign power and of the rhetoric of *p'ungsu*. This evidence helps elucidate Kim's strategy.

What did the civilian president Kim Yŏng-sam intend to realize through such destructive urban rhetoric? What specific narrative of nation would he ultimately challenge and modify? While Kim's spatial politics were generally supported by the prevalent nationalist narrative at a popular level, he explicitly challenged the specific version of official nationalism founded in the former military regime. Here, "official nationalism" is assumed to be the anticommunist, developmentalist, state-sponsored nationalism founded in an explicit form of discourse by the military regime of Park Chung-hee (Pak Chŏng-hŭi). Gi-wook Shin, James Freda, and Gihong Yi (1999) argue that *minjung* (the people's) discourse of the democratization movement in the 1980s and anticommunist and developmentalist discourse of the military regime in 1970 are two confrontational narratives of nationalism in modern Korea. Specifically, anticommunist and developmentalist discourse plays a role as a hegemonic discourse of nation driven by political elites, whereas the *minjung* discourse is constructed as a reactive force against the hegemonic power structure. Both of them have recourse to the rhetoric of nation as the unquestionable national homogeneity.

The destruction of the Government-General Building implies a symbolic disruption of the unstable coexistence of the political and ethnic narratives in the official discourse of nationalism. The ideological mobilization of anticommunist—thus, anti-North—and anti-Japan sentiments at a popular level were actually two sides of the same coin, resulting in a totalizing concept of nation. The Cold War ideology driven by political elites was combined with popular nationalist sentiments founded and promoted by nationalist historiography in the postwar period.[7] Such coexistence was made possible due to the primordialist assumption of the nationalist paradigm: The Korean nation "existed in a 'natural' form," according to a social Darwinist view of historical progress (Shin and Robinson 1999a, 3). In other words, the idea of development, a particular view of modernization, and the primordial concept of nation were interchangeably internalized in Korean nationalist historiography during the postwar era.[8] In the same vein, the cultural nationalists criticized the Chosŏn dynasty for its inability to defend the country (Shin and Robinson 1999b).

The iconoclastic sentiment of the early enlightenment nationalists in Korea was open to contestation. One example is the response of nationalist intellectuals during the destruction of Kwanghwamun (the entrance gate of the Kyŏngbok Palace) in 1926. Kwanghwamun was supposed to be destroyed or at least relocated as part of the construction project of the Government-General Building. *Dongailbo,* a daily newspaper, carried a column written by a reporter, Seol Ui-sik, which personified the gate. The article was full of emotional bitterness, lamenting the fall of the national cultural heritage. In the article, Kwanghwamun was considered as the icon of the Korean nation *(minjok)* rather than the symbol of the former dynasty (de Ceuster 2000, 95). De Ceuster argues that such sentiments reflect the transformation of Kyŏngbok Palace into a national icon in the public's eye. It should also be considered as evidence of the contradictory attitude Koreans had toward the legacy of the former dynasty during the colonial period: On the one hand, placing the blame on the incompetence of the former dynasty was implied in the iconoclastic nationalist discourse; on the other hand, the sentiments reflected in the article imply the recurrent search for Korean uniqueness and a yearning for its past. This ambivalence toward tradition is because of "the manner in which Korea was brought into the modern world" (Robinson 1986, 51).[9] Specifically, the iconoclastic cultural nationalists' rejection of any tradition caused a fundamental dilemma: How to construct a strong national identity without legitimizing their own recent past (Robinson 1986, 50). The intellectual's poetic concern about Kwanghwamun was an implicit expression of a persistent collective desire for tradition as a reference point for the proposed modern nation, which was a missing link in the iconoclastic hypermodern nationalism.

The Western-oriented cultural nationalism of the colonial period paved the way for the realization of the developmentalist nationalism of the Park Jung-hee regime (1961–1979). The self-reliant developmentalism was employed as part of the core political rhetoric promoted by Park's authoritarian regime. Park, the so-called father of the country's modernization, eventually realized the vision of a modernized country that enlightenment thinkers proposed half a century earlier. Taking a similar position to that of the preceding nationalists, fatalism was rejected and the whole society was mobilized under the optimistic slogan of development and achievement through a linear path of progress. Unlike cultural nationalists, Park did not utterly deny the virtue of the past and tradition. Rather, he selectively restructured and reinvented traditional narratives. The anti-Japanese struggle during the colonial period was generally neglected, and national military heroes were discovered from the "distant past" (Oh 1998, 129). Opposing the active introduction of tradition in promoting "national culture," he repressed the succession of traditional elements in social, everyday life, such as fortune telling, shamanism, and *p'ungsu*. The outdated consciousness of the people, assumed to be ridden with premodern customs and practices, was to be reconstructed, allowing for the individual to be born again as a "diligent, self-reliant, and cooperative *(kunmyŏn, chajo, hyŏpdong)*" modern citizen.[10] Therefore, Park searched for the origins of national culture not in the inherited folk beliefs and customs but through the selective reconstruction of upper-class culture.

Through the selective rendering of tradition, the cultural nationalists' dilemma—the contradiction between iconoclastic attitudes toward the past and the search for Korean uniqueness—was ultimately resolved. The Park regime's discourse of national culture provided an ideological background for the modernization of the fatherland. Park's politicized rhetoric of tradition and its successful integration into the modernization project is demonstrated in the contemporary urban landscape of Seoul. For example, a notable statue of historic military hero Yi Sun-shin, built on the main street (Sejongro) in front of Kyŏngbok Palace in Seoul and surrounded by modern skyscrapers, has been considered the paramount symbol of Park's official nationalism, mythologizing the historical figure. In a sense, the "national spirit" represented by Park may be better embodied in the purely modern skyscrapers, a prototype of ahistorical nationality, rather than in historic buildings like Kyŏngbok Palace.

The appearance of the contemporary urban landscape in Seoul shows the contradiction and dilemma of building a modern national identity while retaining traditional structures. The bizarre spatial juxtaposition of the restored Kwanghwamun, the damaged Kyŏngbok Palace, and the Government-General Building in the central area of the capital city had been the representation of the unstable concept of Korean nationhood. This combined landscape of power alludes to a specific discursive field of nationalism: coexistence of Kwanghwamun (official nationalism with a traditional outlook) and the Government-General Building (anticommunism).

The destructive spatial politics of Kim Yŏng-sam's government (1992–1997) inadvertently modified this hegemonic, official narrative of the nation founded by Park. The maintaining presumption of the hegemonic narrative is twofold: (1) the nation as a primordial entity and (2) the congruence of nation and state as political entities. Because of his relative lack of democratic legitimacy, Kim wanted to differentiate himself from the former military governments that sought to erase the colonial memory. When he announced the plan for the demolition of the Government-General Building in 1993, he referred to the need for the "restoration of national spirit." Kim's reference to "national spirit" is not identical with that of Park, even though Kim too mobilized nationalism as political rhetoric. Kim's political rhetoric of nationalism was proposed as a self-sustaining ideology independent from anticommunism and developmentalism. As a result, Park's "political notion of nationalism" (Shin et al. 1999) was critically damaged, and the demolition of the Government-General Building as the symbol of anticommunist South Korean sovereignty provided an opportunity for the new political reference of the nation. In the following sections, I discuss how the controversy ended with the success of the government's plan through popular support and how the naturalized image of the building was reversed in the course of the debate concerning the definition of the nation.

Discursive Contestation: Modifying the Nationalist Narrative of Korea

The heated controversy began in August of 1993, when President Kim Yŏng-sam announced the demolition plan for the old Government-General Building, and it continued even after the dome—the topknot of the building—was eliminated on August 15, 1995. Diverse arguments were made on the fate of the building: prodemolition; conservation; and other negotiating arguments, such as the one from moderates who supported the demolition of the colonial building only after a new museum was constructed. Prodemolitionists included the government, nationalist historians, sociologists, anti-Japanese independence movement–related organizations (for example, Kwangbokhoe, the Independence Restoration Association), and so on. One of the main prodemolitionist figures was Shin Yong-ha, a professor of sociology at Seoul National University. Conservationists were composed of conservative historians and archeologists, architects, some high-ranking ex-government officials, specialists on cultural property and museums, and rightist newspapers, such as *Chosŏn ilbo (Daily Chosŏn),* the newspaper with the largest circulation in Korea.

In fact, many historians and archeologists evidently opposed the hasty implementation of the project because of their concern that it might be harmful to the conservation of properties of the National Museum (Kang 1993, 336). The argument was broadly shared in the field of cultural studies and institutions in terms of the protection of the national cultural heritage. However, the main confrontation was between prodemolitionists and conservationists, establishing a new discursive field of nationalism. Formally, the moderates agreed with the need for the ultimate destruction of the colonial legacy. They only maintained that it needed to be delayed. From the viewpoint of the prodemolitionists, however, the moderates were looked upon as another variation of conservationists in that it would take quite a long time—at least ten years—to design and construct a new national museum and finally move all the cultural properties into it. Therefore, the moderates were ultimately brought into the established discursive field of nationalism on the part of conservationists. Even though they tried to redirect the controversy by presenting another discourse—"conservation of cultural heritage"—this only reinforced the centrality of nationalist narratives by reminding people of the ironic reality of "putting national treasure in the most hated building of the nation" (Pai 2000, 26).

The year 1995 was monumental in the history of Seoul and Korea—it was both the fiftieth anniversary of independence from Japanese rule and the six-hundredth anniversary of the first construction of Kyŏngbok Palace at the beginning of the Chosŏn dynasty. From the view of the government, it was the right time to eliminate the colonial legacy from a space so filled with inherited sacredness. The idea of the demolition of the building was not new. Even in the very first government, there were some arguments for building a new government office, but it was not realized due to the reconstruction cost. In fact, the current demolition project was designed during Roh Tae-woo's term (1988–1992), and President Kim implemented it without hesitation. To help establish legitimacy while succeeding a military government, it was natural for Roh Tae-woo to try to erase the symbol of the former authoritarian regime. In other words, the Government-General Building presented the opportunity to claim historical legitimacy, while at the same time pointing out the weaknesses of the former governments. The issue had remained obscured within the nationalist discourse for the previous fifty years. Ultimately, with the monumentality of the anniversary year, the project was realized as a reform policy to eradicate the remaining vestige of colonial legacy and restore the "national spirit."

The controversy involved identity politics in reading the symbolic meaning of the landscape. Specifically, the critical difference between the two competing arguments lies in identifying contemporary Korean identity with varying historical periods. People read different symbolic meanings of the landscape from different collective memories of different historical eras. For the rightist anticommunists, many of whom cooperated with authoritarian ex-governments (1948–1992), to eliminate the building meant to symbolically negate the authority of the modern history of an anticommunist Korea. For the newly established civil power and nationalists, the building was only a colonial legacy that needed to be removed. Conservationists maintained that the building was a national symbol of modern Korea. Prodemolitionists rejected such ideas.

Hence a discursive field evolved around the attempt to recollect particular narratives of public memory from the substantial history of the building. To the prodemolitionists, such recollection was a discursive strategy to *denaturalize* the landscape. They promoted the demolition project into a nationalist celebration through the constant reiteration of an anti-Japanese, postcolonial narrative of nationalism. For instance, when it was argued that the building was no longer the Government-General Building but the National Museum, the government claimed that "it is wrong to put cultural properties, which are the essence of national culture, into the colonial building" (Kang 1993, 334). Likewise, Shin Young-ha argued that "the social role of the National Central Museum is [to be an] . . . educational institution, . . . encouraging the national esteem for national history and culture. . . . However, the national cultural heritage in the building, the symbol of colonial rule, makes Koreans feel inferior, so we cannot expect any positive educational effect" (1995, 606).

The debate over the various names for the building illustrates the political rendering of historical memory by each group. The *Wolgan Chosŏn* (1995, 389) criticized the government, stating that it wrongfully referred to the building as "the old Government-General Building (Ku Chosŏn Ch'ongdokbu)," noting that the building had been used for the National Museum since 1986. Such contestation over the name of the building is the product of its complex history. The building had four different official names: the Government-General Building (1926–1945); the US Military Government Office (1945–1948); the Capitol Hall (1948–1983); and the National Central Museum (1986–1995). To this list was added at the time of the controversy the *old* Government-General Building. The different names correspond to the different sets of discourse: "the old Government-General Building" directly appeals to the discourse of anti-Japanese nationalism, whereas "the National Museum" is connected to the discourse of the conservation of cultural property. Shin Young-ha emphasized calling it the old Government-General Building and severely criticized those who intentionally used the other name, the National Museum, in the course of the controversy: "They are resorting to every stratagem to conserve the building. They are making propaganda as if the destruction of the old Government-General building is to eliminate the National Museum, and that is critical damage to the cultural property and heritage" (1995, 603).

As seen in this criticism, the struggle to name the building was a product of the discursive strategies for rectifying or delegitimizing the landscape in the public's mind. Prodemolitionists effectively reimposed the social memory onto the landscape through the politics of naming, which assaulted the naturalized image of the landscape as "Capitol Hall." The discourse of cultural heritage was subsumed within the discourse of nationalism sponsored by the prodemolitionists. Accordingly, no interpretation of the building can be entirely free of the main discursive field of nationalism. Thus, conservationists also strategized to refine their argument with nationalist rhetoric. One of the main reasons for conservation was now to serve the needs of younger generations: "In order to avoid repeating the humiliating history, we should examine whether it is better to conserve or destroy it. . . . Thirty-five years of colonial rule are part of our dark history and we should seriously consider how to overcome it and gain instruction from it. Think about why Auschwitz death camp is conserved and whether this case compares to the old Government-General Building" (Lee Sang-hae 1991, 53).

Conservationists went so far as to emphasize that the building was part of the cultural heritage of Korea since it was built by Korean workers with stone from the Korean Peninsula (*Wolgan Chosŏn* 1995). Consequently, their strategies limited the discourses of the two textual communities within the scope of the main discursive field.

Naturalization and the Naturalized Images of the Government-General Building

Four countries have hoisted their national flags above the building through its modern history: Japan (1926–1945), the United States (1945–1948), South Korea (1948–), and North Korea (during the Korean War). This reinforced the contested image of the building. In the postliberation period, however, the image of the building was predominantly identified with the symbolism of the national flag of South Korea. The building was naturalized through its ideological function.

According to James and Nancy Duncan, "the most important role that landscape plays in the social process is ideological, supporting a set of ideas and values, unquestioned assumptions about the way a society *is* or should be organized" (1988, 123). Therefore, they maintain, landscape plays an important role in justifying hegemonic beliefs and values of a society. They describe this process of naturalizing the landscape:

> If landscapes are texts which are read, interpreted according to an ingrained cultural framework of interpretation if they are often read "inattentively" at a practical or non discursive level, then they may be inculcating with a set of notions about how the society is organized: and their readers may be largely unaware of this. If, by being so tangible, so natural, so familiar, the landscape is unquestioned, then such concrete evidence about how society is organized can easily become seen as evidence of how it should or must be organized (Duncan and Duncan 1988, 123).

They introduce several examples to illustrate the naturalizing role of landscape. Drawing upon Nancy Duncan's (1986) study of rich residential landscape in New York, they argue that, judging from the residents' response to the affluent landscape, people "failed to see the relation between the rural landscapes and historic townscapes that they wished to preserve, and the social cost of such preservation," and eventually they could not recognize "the mobilization of bias in the structure of local government, which in effect subsidized the lifestyle of the affluent" (Duncan and Duncan 1988). People responded that "they tended to take it for granted that only a privileged few could or should enjoy the benefits of living in areas of great beauty," regardless of whether they lived in the rich town or other neighboring areas (124). The hegemonic belief structures—"a strong defense of 'home rule' or local determinism and individualistic ideologies"—are concretized in the landscape, and "they are reinforced by reading the landscape" (124).

It cannot be said, however, that every landscape is naturalized effectively to transmit prevailing ideologies. Landscapes tend to "naturalize social relations, though they do not necessarily do so" (Duncan and Duncan 1988, 125). The less textualized, modern urban residential landscape tends to be more easily naturalized because it is usually inattentively and uncritically experienced in everyday life, and the textual basis of the landscape is not clearly known or read. On the other hand, a particular symbolic, monumental landscape, often highly textualized, is more likely to face challenges from contested readings because it is exposed to the critical readings of scholars and specialists, and its symbolic meanings conflict with the everyday, substantial appropriation of the landscape. Some geographical works on public monuments show how they became exposed to the contestation and even subversion of their original meaning (Atkinson and Cosgrove 1998; Osborne 1998; Withers 1996). In these studies, intended meanings in a monument built for the specific purpose of commemoration are easily challenged over time by public perception, as well as in the many works on the erosion and reversion of the original meaning inscribed on a particular landscape. In particular, the Vittorio Emanuelle Monument in Rome shows how this grandiose monumental landscape, which was originally constructed as a national icon in the early twentieth century, became an object of "irreverent amusement and humorous derision" (Atkinson and Cosgrove 1998).

To maintain the monumentality of a symbolic landscape and naturalize it with a particular ideology, some institutionalized devices, performative and visual, need to be arranged. In particular, the Government-General Building had been both a government office and an architectural monument, making it easier to naturalize the building with the prevailing narrative of the nation. During the postwar period, the building became a celebratory locus for the nation as the actual place of the establishment of the first government in South Korea. Through numerous governmental ceremonies in front of the building over the decades and the mass reproduction and proliferation of images from events through the mass media, the colonial building was symbolically transformed into an aspect of the everyday political landscape and the preeminent image of South Korean sovereignty.

Prevailing ideologies were inscribed into the landscape and contributed to the stabilization of the political rhetoric. After liberation, even though a substantial amount of the building's landscape remained the same, the ideology inscribed in the landscape experienced a dramatic transition from colonialism to anticommunism. Crowds of people gathered in front of the building to celebrate the opening of the first government of Korea, which was declared by Lee Sung-man from the second floor balcony of the building on August 15, 1948. The national authority of the landscape was legitimized by attaching the national flag to the center of the building.

In spite of the elimination of the author (the colonial government) of the building, the author's intention, in a sense, was reinforced by a ruling political rhetoric with different ideological contents. Thus the colonial memory of the building was easily eroded and it was replaced with the collective memory of independence and a modern Korea. An image of this event gained naturalizing power by its mass reproduction and circulation. The reproduction was implemented through the official education system, especially in elementary and secondary school textbooks.

Significantly, the new government was not established throughout the Korean Peninsula, but only in the southern part of Korea, a result of the North-South division and partial election. President Lee and the first government were extremely anticommunist, and the transition period, around 1948, was a bloody time of ridding the nation of communists. The image can be identified, therefore, with a particular version of nationalist imaginings: victory of anticommunism over socialism and nationalism in South Korea. It defines the way in which the building is read as the symbol of national independence, the legitimizing icon of the South Korean government, and the symbolic negation of North Korean political power. However, if the landscape is read inattentively, the political, ideological conflicts behind the scenes are easily forgotten. The landscape is uncritically accepted as the pregiven, unquestionable virtue of national sovereignty. The naturalizing force of the landscape comes from its capacity to hide the social history in which the landscape was produced and reproduced.

The imposition of anticommunist rhetoric was more obvious when two Marine Corps soldiers raised the national flag in front of the building after they recaptured it from the North Korean Army during the Korean War. A picture of this event describes the recovery of Seoul from North Korea, representing the building in the backdrop as a sanctified national symbol for South Korea. Through the wide circulation of these visual images, a particular version of national narrative is inscribed on the landscape.

The Role of *P'ungsu* in the Controversy: Revitalizing the Traditional

In the end, the controversy came back to the interpretation of place and space. Prodemolitionists insinuated that the main reason for the destruction of the building was its location in the heart of Seoul. Conservationists aggressively responded that such an argument was nothing but adherence to the unscientific and regressive logic of *p'ungsu*. The controversy eventually led to the division of attitudes toward the premodern past represented in the urban landscape. Shin Yong-ha maintained that Kyŏngbok Palace was the symbol of independent politics and its rebuilding was identified with the modern nationality. Against his fundamentalist position, some preservationists strongly questioned the necessity of restoring the palace to its original form. Lee Sang-hae (1991) argued that the palace was just the symbol of the Chosŏn dynasty, not the icon of modern Korea. According to de Ceuster, there was a critical transition in the symbolism of the palace from dynastic symbol to national icon:

> In a sense, Korea only became a modern nation while under Japanese rule. Although intellectuals had been promoting nationalism since the 1890s, the Korean people only began identifying publicly with their nation during the occupation period. The March First Movement is generally seen as the catalyst in this process of national mobilization. The fate of the palace turned into a metaphor for the fate of the nation. The current reconstruction of the palace can be understood from the same perspective (de Ceuster 2000, 100).

The official event of the demolition shows how the cultural heritage of the past as premodern is negotiated with a modern, popular conception of the nation. On March 1, 1995, before the beginning of the demolition, there were celebratory ceremonial events in front of the building. One important practice, Koyuje, was implemented in the integrated form of official rituals of the royal court and folkloric religious elements. Literally, the goal of the ceremony was to drive out evil spirits associated with the Japanese rule by praying to the gods of Heaven and Land. It was extraordinary that a governmental event was held with such a traditional and folkloric style. By introducing the traditional court ritual, the destruction project was symbolically and visually connected to the reconstruction project of Kyŏngbok Palace. Such elaborately calculated performative rhetoric further shows how government deals with the complex discourses of nationalism.

Whether the demolitionist argument was publicly perceived as *p'ungsu* or not, it was successful. The *Wolgan Chosŏn* attacked the government, asking, "Is this government a *p'ungsu* government?" (Kim Yong-sam 1995). It was a sharp reaction against the government's tacit and enduring appeal to the premodern memory that required the idea of *p'ungsu* as subtext. The controversy was implicitly affected by a prevailing mystical nationalist discourse—"the *p'ungsu* invasion" of Japan on Korean territory during colonial rule. In 1984, some mountaineers and geomancers discovered twenty-five iron stakes on Mt. Pukhansan, which were then interpreted by the mass media as Japan's secret action to sever the flow of ki (vital energy), the spiritual connection of Korean territory. The Government-General Building was regarded as a prime example of the *p'ungsu* invasion by Japan. The building was described as a gigantic iron stake driven into the heart of the Korean territory, which was one of the tactics the *p'ungsu* idea tacitly employed to encourage the nationalists during the controversy.

P'ungsu as a political urban rhetoric was revived during the 1990s and the building's destruction. Even though nobody in the government publicly specified *p'ungsu* elements in the course of the demolition project, the idea of *p'ungsu* hidden within the discourse of "national spirit" played an important role in modifying popular memory and in the imagination of the nation. As the Government-General Building disappeared, citizens looking at the end of the destruction felt relieved to be rid of it, as an iron stake was pulled out (Jung Hye-Seung 1996).

Furthermore, the geomantic imagination was extended to an embodied and sexualized national discourse through the reinterpretation of the traditional *p'ungsu* idea. In the basic idea of *p'ungsu, myŏngdang* (the propitious place) is compared to the human body in Chinese medicine. As a concentration point of the flowing ki, *hyŏl* is the specific point where acupuncture is performed. According to Murayama Chijun (1931, quoted in Kim Sŏng-rye 2000), *myŏngdang* in many cases is the place that looks like or is interpreted as the female genitalia in its topographical appearance. It is no wonder that female sexual organs have been used as symbols of fecundity and richness in traditional agricultural societies. Even in contemporary *p'ungsu,* a propitious place is usually described as the shape of a mother embracing a baby, such as in the works of prominent Korean *p'ungsu* specialist Ch'oi Chang-jo (1997, 187). Therefore, the Government-General Building is reinterpreted as a symbol of the violence (rape) inflicted upon the uterus of the motherland, with the Kyŏngbok Palace being the iron stake driven into the *hyŏl* of the national territory. With these stakes, literally and metaphorically, the national body of Korea is transformed into a sterile female body subjected to violence (Kim Sŏng-rye 2000).

Whether the colonial government intentionally destroyed the Kyŏngbok Palace based on their belief in *p'ungsu* does not concern me. It is certain, however, that the Government-General Building was an explicit challenge to the symbolic geography on which dynastic buildings were located. When Japan decided to construct a new colonial government building, some Japanese architects recommended the site of the Kyŏngsŏngjedae in Daehakro (the predecessor of the present Seoul National University), the foot of Namsan, and the site of the old Kyŏngsŏngbuch'ŏng (Seoul City Hall) (Hŏ 1995). However, upon the colonial government's request, Ito Chuta, a professor at the Tokyo Imperial University, designated the current site for the building and located the Shinto shrine at the foot of Namsan (Bae 2002, 100–101). Therefore, the north-south axis of modern Seoul embodied the spatial connection between political and religious power that replaced *p'ungsu*.

In many aspects, the Government-General Building utterly destroyed the order and ideology represented by the palace. First of all, the new building was overwhelmingly Western and modern in terms of its size, shape, and scale, as well as in terms of its architectural mode. Its vertical edifice was unparalleled at the time. Because buildings taller than royal palaces were considered a symbol of defiance to the king's authority, there were few tall buildings other than the Myongdong Cathedral of 1915 (de Ceuster 2000). The high dome of the colonial headquarters reached 180 feet, its baroque style highly visible from afar, and it was the symbol of a modern and colonial regime borrowing its authority from Western cultural traditions. Some maintain that the building was designed as a reference to the British government-general building in India.

The building was the focal point of the restructured central axis of modern urban structure in Seoul. While the east-west direction was the main axis of traditional spatial structure, newly constructed north-south roads penetrated the urban space. In particular, roads between Whangtohyun (the intersection between Sejong-ro and Jong-ro) and Namdaemun critically transformed the existing urban space by opening up the foreground of the palace (Jung Ki-yong 1994, 52–53). In the traditional spatial structure, the palace was protected from outside invaders by a labyrinthine urban spatial structure. While the Government-General Building was located facing cardinal south, the vertical axis of Kyŏngbok Palace and Kwanghwamun was inclined clockwise by 3.5 degrees (Jung 1994). The difference of 3.5 degrees acutely symbolizes the contrast between modern/abstract geometrical order and topographically modified traditional order.

Conclusion

I have examined how the urban symbolic landscape has been naturalized and focused on the textlike quality of landscape. Specifically, I have shown that multiple readings of the colonial building question the stability of the configuration of national identity in modern Korea. According to Shin, Freda, and Yi (1999), nationalist discourses in contemporary South Korea are described as the confrontation between "anti-communist and developmentalist discourse (state sponsored)" and "anti-American *minjung* discourse." The controversy over the colonial building and its ultimate demolition is the first explicit public symptom of the foreseeable end to such dichotomous opposition, which means that the discursive field of the nation needs to be redefined. Significantly, the emerging national narrative was reinforced by the controversy and by a return to the traditional—especially the nationalist revitalization of *p'ungsu.*

The symbolic landscape not only reflects the official nationalist discourse and naturalizes it, but it also provides a visual and discursive arena where dominant discourse can be contested and transformed. Historically, the space of the Government-General Building had been used to manifest the official political discourse of nation. Through the construction of Kyŏngbok Palace and the colonial building, political elites had intentionally imposed their power and ideology to dominate urban space, and this worked effectively through the processes of naturalization and reification of space. In this manner, landscape was shaped by social discourse and praxis: the product of existing social relations. At the same time, the controversy reveals the active role of landscape in the discursive formation of national identity in Korea. Through the debate, the landscape of power became a communicative device through which the existing official national discourse—the anticommunist and developmentalist narrative—was brought into the realm of public debate. Importantly, the symbolic and representative power of the landscape is not fixed and unitary. This very instability of the represented landscape modifies the virtues that have been taken for granted and the ideologies inscribed onto the landscape.

Notes

1. The Government-General Building of the Chosŏn or Chosŏn Ch'ongdokbu Ch'ŏngsa. The building is located between Kyŏngbok Palace and the front gate, Kwanghwamun.

2. Father of Kojong (1853–1919). He ruled over the country for ten years (1863–1873) instead of his son, because Kojong was so young.

3. After some years of debate concerning its geomantic advantages, Kyŏngbok

Palace was built at the foot of Mt. Pukhansan at the beginning of the Chosŏn dynasty.

4. The popular success of the exhibition can be suggested as another example legitimizing the revisionist critique on the homogenous nationalist narratives in Korea based on the simplistic binary of resistance or collaboration.

5. Many exhibitions at the turn of the century displayed native people from their colonies to entertain citizens of their country and legitimize their colonial policy by showing racial inferiority of the natives (see Grindstaff 1999 for the exhibition of the Philippines at the 1904 Louisiana Purchase Exposition). Japan used the same racial politics of display in many domestic exhibitions at that time. In the fifth domestic Industrial Exposition in Osaka in 1903, Inu and Okinawa natives, Koreans, and Taiwanese were displayed as "living" human specimens.

6. Many Korean intellectuals at the turn of the century had an iconoclastic attitude toward the traditional culture, from Confucian customs to folk beliefs, accepting social Darwinist concepts (Robinson 1988, 28–37).

7. The nationalist historiography in Korea and its anti-Japanese standpoint, officially sanctioned and popularly supported, had existed subordinately to the anticommunist political notion of nation during the period of the authoritarian regime. Therefore, nationalist sentiment was ideologically complementary to the dominant discourse of nation rather than a threat to the ruling power of South Korea.

8. This has been one of the main debates on "colonial modernity" in Korea in recent decades. The key question was concerning how Japanese colonial rule influenced the capitalist-modern development in postwar Korea. Some economic historians argue that Japanese rule critically contributed to modernization of Korea by establishing industrial infrastructure such as railroad systems *(kŭndaehwaron),* but nationalist historians maintain that Japanese rule impeded the internal development of the nation by plundering national resources *(sutalon).* However, more recently the competing standpoints have both been criticized for their uncritical attitude toward modernization and development (Kang 2002). The debate is further evidence that nationalist historiography is not free from a developmentalist paradigm of the nation.

9. "Unlike Japan, the state did not assume the lead in creating the modern nation-state. . . . Korean nationalism did not flow out of a relatively successful challenge to the traditional state from within, as was the case with China. . . . The abrupt failure of the traditional political system helped to discredit the entire Korean tradition" (Robinson 1986, 51).

10. The three slogans of the Saemaul Movement in Park's regime. The movement started as a state-sponsored community development project in 1970, focused on improving living conditions in rural areas. Before long, it was expanded into a nationwide enterprise, with emphasis placed on rebuilding the minds of the people. It was a key ideology for mobilizing people and legitimizing the authoritarian regime.

Part 2

Geographies of the (Imagined) Village

4

Chosŏn Memories

Spectatorship, Ideology, and the Korean Folk Village

TIMOTHY R. TANGHERLINI

AFTER VIEWING a newly opened display of folk materials at the turn of the century, the French journalist Ernest Allard had occasion to write,

> I stood transfixed there, strongly interested, soon quite touched, especially when in passing by the scene I had in front of my eyes, my thought penetrated into the customs of this life of bygone days; because, alas, it appears that the vertiginous evolution of modern progress is making itself felt even . . . [in] those regions where the good old days seemed intent on lasting forever, demolishing in hurried strokes the ancient edifice of costumes and practices, as well as ideas (Allard quoted in Sandberg 1995, 335–336).

Although Allard's impressions stemmed from a visit to the Scandinavian exhibit at the Paris International Exposition in 1878, the remarks could just as easily have been made by a visitor to the Korean pavilion at the Chicago World's Fair in 1893; the Korean Folk Museum when it first opened in postcolonial Korea in 1946; its governmental successor, the Korean National Folklore Museum, when it opened in 1966, when it was relocated to the grounds of Kyŏngbok Palace in 1975, or to its new home in 1993; or Minsokch'on, the Korean Folk Village, on opening day in 1974 (Ha 1980; *Han'guk minsokch'on* 1982 and 1997; Kendall 1999; Kim Youngna 2001; Kungnip Minsok Pangmulgwan 1986a).[1] It probably would not have been written by a visitor to the Lotte World Folklore Museum when it opened in 1989, perched atop the largest department store/shopping mall/amusement park complex in Seoul.

The exhibit at the Parisian fair that inspired Allard to pen his review was designed by the Swedish ethnographer Artur Hazelius, whom many credit with being the originator of the tableau display of folklife in which mimesis and an attendant will to authentic reproduction of the contours of everyday life are the underlying principles.[2] The visitor who views the tableau is invited, often by the accompanying signage, to imagine a story that animates the displayed scene. In the case of Korean folklore museums, these scenes illustrate how particular farming implements or traditional tools were used, freeze moments in time during the performance of rituals from popular religion, or capture small steps in the construction of handmade crafts.[3]

The principles of the museum tableau are brought to their logical conclusion in the open-air museum. Here the cordoned-off display cases of indoor exhibition halls are replaced with entire buildings set on expansive parklike grounds. Actors engaged both in normal activities of daily living and in displays of remarkable individual ability in traditional arts are substituted for the ubiquitous mannequins of the tableau displays. Visitors to the open-air museum are not barred from entering the static story space of the tableau by a cordon or a glass, but rather they walk directly into that space, immersing themselves in a historical kinesthetic experience. This full immersion of the spectator into the historical narrative of the open-air museum heightens the sense of connection with the constructed past and, to a certain extent, intensifies the sense of authentic historical experience proposed by the museum. This "will to authenticity" is an important component of the underlying ideology of the folklore display.[4]

These elements of museum design philosophy surprisingly enough found their way to Pak Chŏng-hŭi's Korea of the early 1970s by way of a rather circuitous path. The first public ethnographic display of things Korean was at the Chicago World's Fair in 1893 (Dilling 1998; Hendry 2000, 54–60; Kim Young-na 2001). During the colonial period, Japanese scholars carried out a great deal of ethnographic and archaeological collection (Pai 2000 and 2001), and subsequent museum displays also included reconstructions of Korean rural life. But these displays were geared primarily toward non-Korean audiences, and they cleaved to the classical view of Korean society as backward, possibly only a few steps removed from primitive society (Hendry 2000, 59). The 1970s folk village, by contrast, developed largely on the path first established by Song Sŏk Ha in prewar Korea and was geared from the very beginning toward Korean visitors. It resulted from grafting the design philosophies of Hazelius' Skansen in Sweden and Olsen's Frilandsmuseum in Denmark onto the burgeoning folklore preservation movement that had lain dormant in Korea since the colonial period.[5] When it opened in the town of Yongin south of Seoul, Minsokch'on perched on the edge of the urban/rural divide. In subsequent years, as urbanization continued apace, Yongin has been effectively swallowed by the outskirts of greater Seoul, and the folk village has been joined by a more recent neighbor, Everland, now the world's eighth-largest theme park, which opened in 1976 as "Yongin Farmland."

The intriguing spatial proximity between the folk village and the theme park is perhaps to be expected, since the outdoor museum and the theme park are closely linked to aspects of industrialization, urbanization, and—ultimately—modernity (MacCannell 1976; Sandberg 1995). Nelson Graburn points out that, in modern society, "the long-held distinction between theme parks and amusement parks on the one hand and museums and cultural centres on the other—between popular and the authoritative—is blurring fast" (1995, 167). Minsokch'on and Everland, located in what was once countryside, perhaps exemplify this blurring best. Now that the greater Everland resort complex includes the HoAm Art Hall, the blurring is essentially complete. Not surprisingly, the two parks are often included on itineraries of domestic and foreign day-trippers and are situated close enough together that even the least demanding tourist itinerary can include both of them in a single day. The visitor can then combine the spatial and temporal fantasies of the two parks—the theme park a seeming flight of fancy into distant worlds and pristine nature and the folk village a voyage into Korea's authentic past.

Most visitors arrive at Minsokch'on seeking a staged experience. Park designers have developed a clear narrative that transports the visitor both through visual and spatial arrangements into this constructed world. In this manner, the intentions of the visitor and those who run the park usually align. My interviews with dozens of folk village visitors over the past seventeen years confirm this: Comments about their experiences at the park at times approach a verbatim repetition of the visitor brochure. Indeed, some have even read to me from the brochure. The structured experiences offered by Minsokch'on and parks like it can accordingly play a significant role in shaping the visitor's understanding of historical and cultural processes. As Dean MacCannell notes in his early yet influential study of tourism in contemporary society, "the best indication of the final victory of modernity over other sociocultural arrangements is not the disappearance of the nonmodern world, but its artificial preservation and reconstruction in modern society" (1976, 8). These reconstructions of the nonmodern world in turn validate the experiences of modern life at the same time as they present a scripted and abbreviated narrative of history that more often than not serves well the ideological agenda of the government or corporation behind the park. Of course, in Korea of the 1970s, there was often little ideological difference between the government and large corporations (Eckert 1991, 258). Nevertheless, the smooth flow of the constructed narrative presented in the space of the folk village is occasionally disrupted by an unexpected event—a mask maker takes off his topknot or a musical performance unintentionally invokes a political message. It is during these usually brief interruptions that the contours of the park narrative are laid bare and its political nature emerges.

At Minsokch'on, the visitor is encouraged to wander back in time into a painstakingly reconstructed space—a reconstruction based on the park designers' and conservators' conceptions of folklife and history. Benedict Anderson, in *Imagined Communities* (1983/1991), proposes that the museum—and by extension, the open-air museum—is intimately linked to the processes of "imagining the nation." He notes that "museums, and the museumizing imagination, are both profoundly political" (178), and he suggests that "as with modern persons, so it is with nations. Awareness of being, imbedded in secular, serial time, with all its implications of continuity, yet of 'forgetting' the experience of this continuity—product of the ruptures of the late eighteenth century—engenders the need for a narrative of 'identity'" (205). It is precisely this type of narrative of "identity"—a narrative that is deeply political since it is tied intimately to the official projections of the national narrative of origins, heritage, memory, and the past—that inspires cultural officers of the Pak regime to rediscover the colonial-era fascination with Korean archaeology and folklore. The ruptures that Anderson mentions need not be traced as far back as the eighteenth century for Korea; in fact, the ruptures of the last hundred years are more than adequate to have provoked this response (Kendall 1999).

Perhaps to better understand the motivations for building the folk village, it may be best to take a short detour through the development in Korea of folklore museums and folk villages and the attendant development of folklorists' conceptions of rural life, since the conception of the "folk" in Korean folkloristics is almost exclusively rural (Ch'oe In-hak 1995; Im 1991; In 1978; Kim Son-p'ung 1994; Kim Tong-uk et al. 1988). Roger Janelli (1988) notes that the study of folklore in Korea is nationalistic, by which should be understood that Korean folklorists focus primarily on the cultural expressions of Koreans; and historical, by which should be understood that Korean folklorists tend to consider the development of culture over time and to seek out the earliest examples of any given expression. In the following short narrative of the development of folklore studies in Korea, I want to highlight how the field developed into one emphasizing the unique cultural expressions of the Korean people, predicated on a master narrative of "5,000 years of Korean history." Folklorists have, as the field developed, settled on situating these expressions of "Koreanness" historically in the late Chosŏn and placing them geographically in the countryside (Kim Son-p'ung 1994; Kim Tong-uk et al. 1988). In turn, this Korean answer to Alan Dundes' (1977) question, "Who are the folk?"—or perhaps better expressed in Korean, "Who are the *min?*"—explains to a great extent the formal features of Minsokch'on and the other Korean folk museums.

Folklore study in Korea made its first great strides paradoxically during the colonial period (In 1978; Janelli 1988). Although the early Japanese policy was one of cultural erasure, as the colonial period wore on, policies changed so that the study of Korean culture and history were encouraged, albeit with an implicit goal on the part of the colonial authorities to emphasize not only a long connection between the cultures of Japan and Korea, but also to show the cultural reliance of Korea on Japan. An indirect result of these changing policies was the emergence of cultural nationalism—*munhwa minjokjŭui*—a trend described well by Michael Robinson (1988). As Robinson has noted, with cultural nationalism came an increased focus on Korean history and culture by Korean scholars. Roger Janelli (1988) situates the beginnings of folklore as a field of scholarly inquiry in Korea during this period, a view echoed by Ch'oe In Hak (1995, 20). Supporting the views of In Kwŏn-han (1978), Ch'oe further links the emergence of folkloristics—*minsokhak*—to the end of the *silhak* (practical learning) period, opining that two of the last *silhak* scholars, Yi Nŭng-hwa and Ch'oe Nam-sŏn, were also the first modern folklorists in Korea (20–21).

Yi and Ch'oe, however, were not alone in their folkloric endeavors during the colonial period. Most overviews of the history of Korean folkloristics mention two other prominent scholars—Son Chin-t'ae and Song Sŏk-ha—who stand out, albeit for significantly different reasons. Son saw the study of folklore as closely related to the study of the life and hard times of the disenfranchised masses and hypothesized an early, neocommunist state situated on the Korean Peninsula (Janelli 1988). The Marxist slant of his scholarship prefigures the emergence of the *minjungjuŭi* discourse that informs Korean folkloristics during the middle to late 1980s. Song Sŏk-ha was perhaps the most contemporaneously minded of these early Korean folklorists, undertaking a great deal of fieldwork himself (Chang 1991; Kungnip Minsok Pangmulgwan 1996). He is perhaps best known for founding the first Korean folklore museum, the Kungnip Minjok Pangmulgwan in 1946. Rather than using his studies of folklore solely as a form of resistance to the Japanese colonial presence as most cultural nationalists did, Song also took a page from the book of the European Romantic nationalists. Like his Swedish counterpart Hazelius, Song's conception of the folk was firmly rooted in the peasantry, and his later museum expressed nostalgia for what he perceived to be a vanishing rural culture. The period that Song envisioned as the location of Korea's national cultural origins did not hearken back to the mythical times that Ch'oe Nam-sŏn and Yi Nŭng-hwa attempted to recuperate in their mythological studies but rather to the eighteenth and nineteenth centuries. Song was a preservationist, a position inspired by European folklorists from the nineteenth century like Hazelius, who felt that industrialization and modernization threatened the rural lifestyle that provided the cultural context for the folk expressions of national spirit (Chang 1991; Song Sŏk-ha 1934a).

Song's connection to Europe and European cultural display was by no means as vicarious as one might initially suspect. After his college years in Japan, he spent a great deal of time and energy working for the internationalization of Korean folklore, and he tried to develop close contacts with European folklorists and folklore societies (Chang 1991). During the mid-1930s, one of his closest international contacts was the Swedish adventurer and zoologist Sten Bergman, to whom he eventually gave copies of his films of mask dance drama to show in Sweden. Bergman had been sent to Korea by the Swedish Natural History Museum to collect specimens of birds and animals and by the Swedish Ethnographical Museum to collect aspects of Korean folk culture. In his book, Bergman (1937, 221–222) recounts a somewhat surprising event in which one of his Japanese hosts has taken a photograph of Bergman's children at Skansen: "Just about this time a very curious thing happened. My children, during my absence from Sweden, were taken one day to Skansen. . . . They were wearing Swedish national costumes and for this reason had caught the eye of a Japanese visitor there. He had asked them permission to take a photograph of them and with characteristic Japanese courtesy had promised to send them a copy. . . . The Japanese added his address, which turned out to be at Keijo (Seoul)." Bergman goes on to visit the Japanese family and receives a copy of the photograph. This odd coincidence brings Skansen that much closer to Korea and into the immediate realm of Song, who at the time was developing his ideas for the Korean folk museum.

At the same time, Song's close colleague and cofounder of the folklore society, Jŏng In-sŏp, traveled in 1936 to Copenhagen where he not only attended the congress of the Comité International Permanent des Linguistes but also visited various museums, probably including Olsen's Frilandsmuseum (Yun Hŭng-no 1991). Of course, I do not wish to suggest that Song's sole inspiration for creating his museum came from these Scandinavian contacts but rather to underscore that the museum design philosophies of Song and his later counterparts were not solely based on Japanese models; they rather included a clear appreciation of developments in Europe—in particular Scandinavia. Of course, as Hendry (2000, 7) notes, many of the early Japanese ethnographic displays had these same Scandinavian models as one of their sources of inspiration. Korean folklorists in the 1960s clearly took Song's lead, positing the Korean folk as a rural populace and historically situating them in the late Chosŏn. In this manner, the rural population of that period came to be interpreted as the historical and geographic locus of the unique Korean national spirit and became the focus of their folklore displays.

Because of Song's rural nostalgia, his 1946 folklore museum became a repository for items that were related to the vanishing agrarian lifestyle

(Kungnip Minsok Pangmulgwan 1996). This museumification of nineteenth-century rural culture was started already in 1893 when, at the Columbian Exhibition in Chicago, the Koreans exhibited "a seven-room house in traditional style, and a display of apparel, eating utensils, [and] furniture" (Dilling 1998; Kim Young-na 2001). Song's museum offered the opportunity to imagine a story of rural life through the presentation of artifacts of folklife in idealized yet typical situations. In viewing the items displayed at the museum, which were often arranged in tableaus as was the 1893 exhibition, urban spectators could recover a lost dimension of Koreanness—or more accurately, late Chosŏnness—that was the wellspring of their collective identity, regardless of whether they had ever had those rural experiences. This early museum thus became a site for the recovery of lost memories that never were, and in this sense it began the process for the predominantly urban visitors of the folklore museum of imagining a Korean identity that was closely linked to an agrarian past and situated both chronologically and geographically outside the realm of their own experiences.

Korean folklore endeavors came to a screeching halt with the advent of World War II and did not begin to revive until the late 1950s (Ch'oe 1995, 21–22). In the early 1950s, during Yi Sŭng-man's reign, the government launched widescale literacy and antisuperstition campaigns. While these policies were intended to help modernize the country, the results for folk tradition—and, by extension, folklorists—were catastrophic. Shamans were driven underground with an attendant decrease in customers, people abandoned aspects of their ritual life that provided a strong connection to both family and community, and folk belief was seen as something that should be eradicated. Ironically, at least in the realm of folklore, the differences between the early Japanese policies of cultural erasure and the antisuperstition campaigns of Yi Sŭng-man are, in retrospect, somewhat difficult to discern.

After the abbreviated administration of Yun Po-sŏn, Pak Chŏng-hŭi assumed power and, with a heavy hand, vowed to move Korea into a position of global industrial prominence. Among the most influential of his policies in the rural areas were the National Reconstruction Movement for rural development begun in 1961 and the New Village Movement (Saemaŭl Undong) launched in early 1971. Whereas the antisuperstition campaigns of Yi Sŭng-man had been aimed at aspects of day-to-day life, these new rural initiatives set their sights not only on the "backward mindset" of the rural populace but also on the physical representations of rural culture. As Ch'oe In-Hak notes, one of the government's main goals in the 1960s and 1970s was the physical modernization of the countryside—straw roofs were replaced by zinc ones, dirt roads were paved with asphalt, mud walls were replaced with cement blocks. Even *changsŭng,* poles that housed guardian spirits and stood watch over the entrance to the villages, were uprooted and destroyed (Ch'oe 1995, 24). By the early 1970s, these once ubiquitous *changsŭng* had all but disappeared from the landscape.

But a funny thing happened as Pak's modernizing agenda entered its second decade. While the regime was hell-bent on industrializing the country, they also had a vested interest in presenting a unified narrative of national identity that would help legitimize their authority. Although the government's intended representation of Korea was certainly supposed to be one of a forward-looking country, it also wanted to project a sense of a Korea that had a long, impressive, and distinct history, whose people expressed themselves in unique and noteworthy ways, and that was unified under a strong central authority. The rural development initiatives had been remarkably successful not only in modernizing the Korean countryside but also in wiping out significant aspects of rural culture—a culture that had been linked by Korean folklorists such as Yi, Ch'oe, Son, Jŏng, and Song, as well as their modern counterparts to the spirit of the nation. As a result, numerous cultural initiatives were intensified in the 1970s to collect and preserve aspects of Korean culture that were threatened by the Korean state itself, thus echoing Song Sŏk-ha's preservationist admonition from the 1930s, in the first volume of *Chosŏn minsok,* where he laments, "The materials of our unique folk culture are disappearing one by one" (Janelli 1988, 40; Song 1934b).

Although the Law for the Cultural Properties Preservation, modeled on a similar yet earlier Japanese law, had been promulgated early on in the Pak regime, the enforcement of the law during the 1960s tended more toward the preservation of items of great artistic, archaeological, and historical significance (Pai 2001). Indeed, most of the early efforts of the Munhwachae Kwalliguk (Bureau of Cultural Property Preservation) were focused on identifying archaeological and artistic national treasures *(kukpo).*[6] Less importance was attached to elements of folk culture. While certain folk expressions came to be designated as "intangible cultural properties" and some of the practitioners dubbed "living cultural treasures," most aspects of folklore and folklife were either consciously obliterated or, in the best cases, simply ignored during the early years of the Pak regime. By the late 1960s and early 1970s, however, a sense of impending cultural doom coupled to Pak's own presentation of himself as a man of the people—someone not averse to rolling up his pants and getting his feet dirty during the rice transplanting—gave impetus to a wave of preservationist activities in the realm of folklore and folklife. It is not surprising then, that—along with extraordinary oral literature collection efforts such as the one started in 1968 and finally presented in the eighty-four volumes of the *Kubi munhak taekye*—the designation of folklore items as important national treasures and a general wave of folkloric collecting reached full tilt. By way of illustration of this increased focus by the Munhwachae Kwallikuk on folklore starting in the late 1960s, only seventeen folkloric performances were designated as intangible cultural properties from the time that the law was implemented through 1968. In the following two years, however, fifteen more folkloric events received this designation, and from 1969 through 1973, a total of twenty-nine were added—more than one-fifth the current total of 109 (see Table 4.1). It was during this time of attention to folklore that Minsokch'on came into being.

Originally funded by the Kihŭng Tourism Co., Ltd. (later the Chowŏn Tourism Promotion Co.), construction of Minsokch'on had enthusiastic government backing and a mandate to preserve and present the "quickly disappearing" folk culture of Korea (*Han'guk minsokch'on* 1980). The folk village is described in a recent tourist brochure as "a living museum that recreates the lifestyle of [one or two] centuries ago. There are potters, weavers, blacksmiths, and other artisans who practice their trades in traditional fashion. There are also two hundred and forty traditional homes and a small amphitheater for music and folk dances" (*Han'guk minsokch'on* 1997). The reproduction of life of a hundred or so years ago at the village positions Korean folklife as a feature of the late Chosŏn period—the "good old days" of Allard's review—and begins the process of both imagining the source of a Korean identity that is far enough removed from the present to be a forgotten past, yet still close enough to the present time to be remembered. It

TABLE 4.1. National treasures and cultural properties, 1997 and 2004

Classfication	Total 1997	Total 2004
National treasures	293	306
Treasures	1,232	1,401
Historic sites	384	452
Historic and scenic sites	6	9
Scenic sites	7	12
Natural monuments	286	436
Important intangible cultural properties	103	116
Persons with skill	181	367
Important folklore materials	228	244
Tangible cultural properties	1,619	2,256
Intangible cultural properties	193	327
Monuments	1,053	1,583
Folklore materials	266	319
Cultural properties materials	1,435	1,999
Traditional architecture	21	—
Preservation areas	2	—

Source: Munhwajaekwallikuk Statistical Overview (1997; updated 2004).

is a profoundly ironic temporal positioning, since the nineteenth century was hardly a period of peace and serenity but rather a time marked by internal political difficulties, including several peasant revolts, social strife, and crises of international scale—a past that is clearly and deliberately forgotten in the context of the village.[7]

A visitor to the folk village enters the sprawling complex through a large three-passage gate *(sammun)* and is thus transported back to an imaginary village of the nineteenth century. A bit to the left and beyond a collection of sufficiently frightening *changsŭng* lies the Filial Son Gate (Hyojamun), a small monument dedicated to the filial acts of Yi Tok-kyu (1850–1900) built in 1904.[8] Imbued with the proper sense of filial devotion and, by extension, a sense of deference to authority, the visitor continues on a deliberately indirect path that leads past replicas of farmhouses from the entire peninsula, small fields planted with typical crops, the mansion of a *yangban* (noble) family, a functioning Buddhist temple, and a government office, complete with jail and torture devices.

Along the meandering path that winds its way through the village and passes by the numerous houses that dot the park grounds, the visitor can stop to listen to a raucous performance of farmer's band music *(nong'ak)*, perhaps pass by a group playing on a seesaw, or even watch a performance of tightrope walking. The visitor can also stop in at the fortune-teller's house and buy a charm to ward off evil spirits, pop in to buy a traditional pipe at the pipe maker's workshop, or stop to watch mulberry paper being made at the paper mill. After this promenade back through time, the visitor can wander into the marketplace *(changt'ŏ)*, where artisans in small booths produce handicrafts such as masks and musical instruments and *hanbok* (traditionally) clad women sell trinkets for shockingly contemporary prices. Before finally heading home, the weary visitor can sit back on a raised platform and quaff one of the many folk drinks *(minsokju)* offered at small stalls around the marketplace's perimeter. The "proper" route through the folk village is made explicit through signage, visitor maps, and—in a recent development that reveals the penetration of the mediascape into this invented ethnoscape—with an elaborate "cybertour" that proposes four possible itineraries through the park grounds: an introductory route, a route for families with children, an exhaustive route for those who arrive early, and a quick, "greatest hits" tour for those who arrive later in the afternoon (Fig. 4.1).

The folk village is not only a fiction of time and history, it is also a fiction of geography. Although tourist brochures and the museum catalog stake claims to authenticity for the village, Minsokch'on is not in fact a true village but rather an imaginary landscape, despite the wishful thinking of the catalog, which calls the park "a replica of a typical nineteenth century Korean village" that captures an "authentic atmosphere" (*Han'guk minsokch'on* 1982). Unlike other folk villages, such as Ha Hoe village,

Songŭp, or Yangdong, villages that existed long before receiving the designation of "folk village" by the Munhwaje Kwallikuk in 1984, Minsokch'on has no such historical pedigree (Moon 2001). Its position as tourist site is not a reinscription of a lived space.

Instead, while the catalog says that the museum is laid out like a regular Korean village, it in fact does not mimic in organization, building style, or any other readily discernible geographic features the patterns of an actual village (*Han'guk minsokch'on* 1980 and 1997). The buildings are all in pristine condition—thatched roofs are full and thick, *ondol* floors are perfectly sealed, windows are square and walls are plumb. The houses sit much farther apart from each other than would be expected—or perhaps far too close together by many hundreds of kilometers—and there is no sense of a village center point or hamlet divisions. A water mill straddles a stream that runs in an unlikely relationship to nearby structures, and farmhouse courtyards and stalls are kept immaculate and free of animals. The Buddhist temple crowds the unusually situated regional government offices, and a small farmer's cottage nestles up against the walls of a *yangban* estate. Although several of the buildings at the folk village were taken timber by timber from various parts of the southern Korean Peninsula and painstakingly reassembled in situ, as in the case of the Hyojamun, the majority, such as the government office and the Confucian academy, were built with new materials from old plans or descriptions.[9]

Unlike the Cheju folk village, another constructed landscape but one

FIG. 4.1. Cybertour (from http://www.koreanfolk.co.kr).

that actually succeeds in mimicking the spatial organization of a Cheju village because of its intense localization—a peon as it were to the fierce regionalisms that characterize internal Korean politics and that resist the hermetic closure of the nation—Minsokch'on attempts to act as a representation of all of Korea (a house from Chejudo sits unabashedly next to a house from Ch'ungch'ŏngbukdo, for example), even while insisting on its status as authentic replica. The regionality of a museum such as the Cheju folk village is an impossibility for a museum such as the Minsokch'on because, unlike the Cheju village, Minsokch'on does not attempt to represent Korean life as it was lived in a specific place. Rather the museum strives to represent Korean life as it was never lived—and situated in no particular place at all—a narrative strategy reminiscent of the fairy tale (in this context, it is interesting to note that Korean fairy tales frequently open with the formula, "Yetnal yetnal e, Chosŏn sidae e . . ." [A long, long time ago, in the Chosŏn period . . .]). Paradoxically, this nostalgia for a time that never was, marked by an anonymity of place and persons who never were—a nostalgia that Appadurai (1996, 77) aptly labels "imagined nostalgia"—is what imbues Minsokch'on with the power to emerge as part of the narrative of national identity that lies at the root of the motivation for its initial building.

The original audience for the folk village was largely Korean, and it seems likely that visitors to the museum in the early 1970s did indeed interpret the village in the context of the government narrative of national identity. In fact, a large percentage of the early visitors to the village were schoolchildren on field trips. As Shin Gi-wook (1998) has pointed out, in the 1960s and 1970s the educational system was mobilized by Pak's government to inculcate young Koreans not only with a narrative of 5,000 years of Korean history but also 5,000 years of strong Korean government. A visit to Minsokch'on played into these educational goals. Here, schoolchildren had an opportunity to see the peaceful rural past of their now modern Korea, where farmers wearing clean linen *hanbok* lived in simple yet exquisitely crafted houses, worked in perfectly tended fields, and had plenty of spare time to play farmer's band music, bob up and down on seesaws, or watch a tightrope performance. As the schoolchildren moved through the park, they would also gain an understanding of the grandeur of the nobility *(yangban)*, the importance of education based on the tenets of the Confucian academy, and the fair but firm hand of the government. By the end of their visit, the children would have gained an understanding of Korea's long, unique, and idyllic past, as well as an appreciation of the benefits of capitalism, industrialization, and urbanization.

The folk village also offered new urban dwellers a glimpse of the rural life they had recently left behind, and it perhaps could play into a nostalgia for a country idyll abandoned in favor of the opportunities of the city. With a visit to the village, the urban dweller returns to an anonymous yet conveniently located home town *(kohyang)* and experiences once again the rhythms of the countryside, devoid of the noisome aspects that accompany actual rural life. Walking through the three-passage gate, the visitor is transported into the distant world of their great-grandfather's youth, where fields of rice glint green on a hazy summer day and storage jars catch the long rays of the setting sun. Visitors can thus reminisce about their own rural past, regardless of whether they had ever experienced it. Interestingly, this narrative of reminiscence also allows for a reversal of the rural idyll: "Think how much better we have it now than we did back then!" the visitor can muse. This new reading, in turn, validates the breakneck speed of South Korea's industrialization.

The Danish folklorist Michael Chesnutt (1999) proposes that folklore emerges from the dialectic tension that exists between the individual and tradition. The representations of tradition inherent in any folk cultural display force individual visitors to that display into a position in which they necessarily interrogate not only their own relationship to tradition but also the extent to which they can accept the narrative of tradition presented in that environment. Sandberg, in his examination of the folk museum in late-nineteenth- and early-twentieth-century Scandinavia, proposes the theoretically rich notion of "spectatorship," a position that invests the visitor to a museum space with individual agency. Spectatorship stands in marked opposition to the concept of display, a concept that evacuates this positionality of agency and substitutes a more static consideration of objectivity. In his study, Sandberg proposes that the museum visitor, the spectator, enjoys the liminality, the in-betweenness, inherent in the potentially contradictory position of fully entering a space that is narratively defined as traditional—as is the case at Minsokch'on—and the individual's own ongoing negotiation of the traditional in everyday life. Because of the tension that arises in this confrontation of the individual and the traditional at the site of the folk village, the role of the folklorist in the study of such places takes on an added dimension. Not only does the folklorist need to map the contours and the historical basis of the narrative of tradition presented by the museum, but he or she must also explore the individual responses to that narrative.

Despite what the folk village curators might want, the museum visitors do not necessarily reconcile their own relationship to the traditional with the projection presented by the village. Rather, the individual visitor engages the space with the understanding in mind that the environment, while seductive, is a construct. Even though the "traditional" presented in these places offers a seemingly "thicker" narrative of identity-as-heritage than perhaps is found in everyday life—an ethnographic thickness that many of my informants cite as the initial impetus for them to seek out Minsokch'on—once at the folk village, the visitor rarely accepts uncritically the proposed narrative. Instead, I have found that the individual's engagement with the proffered narrative of tradition is one marked by sliding interpretive registers on a grid of understanding anchored at its extremes by a series of nodes. As a form of shorthand, I label these nodes the complicit, the compliant, the resistant, and the ludic. These positions are influenced by the visitor's varying degrees of experience with rural life and knowledge of history, and they range from the completely uninformed to the very informed. Similar to the individual's negotiation of the traditional in everyday life, no single museumgoer ever occupies any of these extreme points in his or her interpretive engagement with the museum. Instead, individual interactions with the "traditional" as presented at the museum cluster around the nodes; neither the nodes nor individuals' interpretive stances are static. Rather, both are mobile and set against a fluid, rapidly changing social and political background, where the meanings of the nodes as ideal categories and the individual's gravitation toward these nodes are in a constant state of flux, even during their visit to the park.[10]

To illustrate briefly the sliding interpretive registers that visitors engage on a visit to the folk village, I offer several of my key informants: Kim Min Sun, a researcher with the Cultural Properties Preservation Research Institute; Pak Rae Hwan, a mask maker and sculptor who used his employment at the folk village as a means to fund his burgeoning artistic career; Kim Yŏng Hwan, an international studies student at Yonsei University in the late 1980s; and Stephen Coleman, a young American who taught English in Seoul for the summer of 1999. Perhaps I could also have offered an elementary school tour group and their teacher and myself as examples, but I believe that in the first case, the positionality of the group teeters on the edge between the complicit (the teacher) and the compliant (the students) and a naive form of the ludic (seen in the playful use of the torture bench at the district governor's office); as for myself, while allowing for a tempting descent into the self-reflexive, my interpretive engagement with Minsokch'on is conflicted and constantly changing—and also informed by my own development as a folklorist with experience both in Korean villages and cities, as a student intrigued by progressive politics at a Korean university in the late 1980s, as a junior researcher working for a large Korean governmental institution while being supported by a foundation whose fortunes are closely tied to the ever-expanding global mediascape, and later as an American academic. A similar complex series of engagements would obtain for my colleague, a well-known scholar of Korean Christianity, who recently held his wedding at the folk village; perhaps these last two cases are illustrative, as they belie the seeming simplicity of interpretation that mention of Minsokch'on at times evokes. Indeed, one of my colleagues at Yonsei University, a sociologist, was recently puzzled by my interest in the folk village, saying, "Is that really important?"

I first visited the folk village with Kim Min Sun, at the time a curator at the National Folklore Museum, several weeks after my arrival in Korea in June 1987 as part of my crash course in Korean folklore and folklife. She has been a gracious host to the village on five occasions since then. Through her work at the folklore museum as it prepared for the Olympics, she had become attuned not only to the general theory of cultural display but also to the specifics of display and performance at the folk village—indeed, since she knows of my familiarity with Scandinavian outdoor museums, she has over the years been particularly interested in learning how "successful" the folk village is in comparison to its Nordic ancestors, Skansen, Maihaugen, and Frilandsmuseet.

The route we have followed on our first and subsequent visits to the folk museum has always been the same—although as the folk museum has been built up, it has increased somewhat in length—and has always hewn to the prescribed pathways through the village. The discussions of the buildings and the various performances and other aspects of folklife presented at the village have also closely followed the village map, although—given her specific expertise in folklife and my own deepening knowledge over the past seventeen years—the discussions have become more detailed and our excursions perhaps more leisurely, lingering over a newly acquired money chest here or marveling at the skill of an acrobat there. Despite these discussions of form, Min Sun has never once questioned the "authenticity" of the space, despite my queries about the seeming impossibility of an Ullŭngdo Island house perched next to a pond and a house from Chungchŏngpukdo, or the obvious elision of regional differences in the farmer's band performance, where Chwado and Udo styles commingle unabashedly. The day with Min Sun always ends with a visit to the marketplace, usually for lunch, and then a quick detour over a footbridge to avoid the increasingly large amusement park built to compete however meekly with the nearby megapark of Everland. In our coengagement with the folk village, there are few questions about the adequacy of the narrative, nor do questions arise as to why a vision of rural late Chosŏn quiescent to the demands of competing regionalisms and seemingly oblivious to the stress of industrialization and the encroachments of globalization should be presented as the site of an originary narrative predicated on the cultural nationalism of early scholars such as Yi, Ch'oe, and Song. Instead, "Here is what makes Korea Korean" is the repeated implicit message of all these visits.

While the visits with Kim Min Sun are the closest I have ever gotten to outright complicity with the narrative of the park, the visits I have made with numerous tourists, domestic and foreign, and the interviews I have conducted with them suggest that many of the visitors are utterly compliant with the proposed narrative and the proposed route. Stephen Coleman, who had been in Korea for only two weeks when I met him, found the folk village to be a "pleasant respite" from the frenetic pace of Seoul and mentioned that the folk village allowed him to "get a feel for the real Korea." Accompanying him through the folk village grounds, I was impressed by his tenaciousness in sticking to the prescribed pathways and his willing engagement with the park narrative, despite the awkward translations in the brochure—his admission of an unwillingness to travel too far afield in Korea ("too much of a hassle") stood in contrast to the seeming adventuresome nature that had brought him to Korea in the first place. His engagement with the park was not too removed from that of countless foreign tour groups (admittedly, I was a bit surprised that he was not part of an organized tour) bused in and accompanied through the grounds by flag-waving guides, nor to a certain extent was it removed from the parade of schoolchildren, who still constitute the largest single group of visitors to the park. That is not to say that the visit was as serious an affair as my at times dour walks with Kim Min Sun. Rather, while there were moments of playfulness, moments of enjoyment, and moments of introspection, these moments were all contained within the bounds of the scripted environment. A walk up the hill to the monastery encourages introspection, while attending a farmer's band performance provides musical entertainment—and perhaps even a little dancing. A visit to the *yot* (taffy) maker encourages similar enjoyment and joking, particularly among students, while the torture rack begs to be tried. Similarly, the marketplace and food stalls all present opportunities for scripted play. But none of this play rises to the level of the carnivalesque, the ludic, that also stands as an extreme positionality.

By way of contrast, the ludic engagement with the space was immediately evident in mask-maker Pak Rae Hwan's insider maneuverings through the terrain. On one of my many visits with Rae Hwan, I was captivated by his movements through the park and his easy flow in and out of character. One moment he would pose for a picture, his topknot perched smartly atop his head and his "work clothes" completing the image of the rural idyll many visitors had come to experience. The next moment he would remove the hat and guide me along the closed alleyways between buildings and farm enclosures. Understandably, he paid no attention to park signage and moved through, across, and in between the otherwise well-structured spaces. Indeed, our movements through the park seemed to me utterly chaotic at first, providing unexpected views of buildings and people. Occasionally, on my first whirlwind tour of the park from his perspective, he would stop to point out some aberration or to comment on the arrival of a new building, or he would stop to flirt with his fiancée selling trinkets in one of the booths, scurry off to introduce me to his friend the pipe maker, where we talked about the upcoming presidential elections, or duck into a disguised employee rest area for a smoke and a chat with friends. On one particular afternoon, while on the way back to his mask-carving shop, he caught the eye of one of his friends, a tightrope walker also on a break, and they pretended for a moment to be normal park visitors. To the bemusement of other visitors, the two proceeded to launch each other several meters into the air on a seesaw, an impossibility for the untrained. For Rae Hwan, the folk village stood as little more than an elaborate playground—his movement through the space was playful and contestatory, completely ignoring the park's narrative of national identity and unity. The folk village, however, was also his place of employment and where he sold his masks to visitors from all over the globe (indeed, shortly before the 1988 Olympics, I spent one evening helping him write signs with prices in dollars and yen as well as wŏn: "Yangban mask, $29.95"), and because of his commercial investment in the space of the folk village, he was on some level entirely implicated—complicit as it were—in the manufacture and sale of the museum's constructed narrative of cultural identity, even while he was engaged in his own playful refiguration of that same narrative.

Although Rae Hwan's ludic engagement with the park offers its own form of resistance, it is a resistance qualitatively different from the resistance I encountered in my visit with Kim Yŏng Hwan in late November 1987. While Rae Hwan's main resistance was physical, contesting the boundaries of the park, Yŏng Hwan's resistance was primarily ideological. Indeed, the well-orchestrated narrative of national spirit carefully constructed in Minsokch'on experienced a significant interruption at the beginning of the Chŏn Du-hwan era. A crisis of trust in government accompanied the Kwangju massacre of 1980 and the martial rule of the first fifteen months of Chŏn's regime. A coordinated spirit of resistance arose and was bolstered by the increased emphasis among dissidents of the *minjungjuŭi* concept. Thus, in the 1980s, while the intended narrative of Minsokch'on certainly continued to be one of the main interpretive modes available to the visitor, the potential for a narrative of resistance emerged alongside this official narrative.

I met Yŏng Hwan in a *p'ungmul* (farmer's band music) class in which I had enrolled, both to learn more about Korean traditional rhythms and to get entry into the student-dominated world of the democracy movements. Our *p'ungmul* lessons came with a heavy dose of progressive and, at the time, clandestine Korean history, coupled to readings and discussions of Marx and revolution. Initially, I was surprised when Yŏng Hwan and a friend of his proposed a visit to the folk village, after our group had played at the beginning of one of a long series of protests at Yonsei University leading up to the presidential elections slated for late December. Partly, he and his friend wanted to listen to the farmer's band music groups that had been invited to play that weekend at the folk village, but they also wanted to introduce me to "*uri nara*" (our country). Not surprisingly, the majority of our visit was split between the performance amphitheater listening to the bands and, being students, the market, where I was given lessons in the regional and class differences of a wide array of alcoholic drinks. However, the beating of the *chang'gu,* the hourglass drum, in the farmer's band performance unexpectedly acted as a powerful disruption of the well-crafted park narrative, forging a direct yet unintentional link to the student movement, political protests, and *minjung* discourse (Tangherlini 1998). Immediately, the *yangban* mansion was cast in the light of class struggle and the small houses were seen as honest representations of the life and hard times of the rural poor—the encounter with this differential in land apportionment led to a discussion of the contemporaneous rampant real estate speculation that was chasing people from their houses and bulldozing communities. The houses reconstructed from the North were looked at longingly in the context of national reunification, while the regional government office became part of a discussion of police, surveillance, and torture, and thus aligned well with our experiences the previous weekend at the student demonstration. The shaman's tree was seen in the context of indigenous religious movements and the resistive nature of shaman ritual that Kim Kwang-ŏk (1994b), Choi Chungmoo (1993), and Ryu Je-hun (this volume) have explored, and the Buddhist monastery was seen in the context not only of government policies that forced Buddhist temples into the hills during the Chosŏn dynasty but also the encroachment of Christianity.[11] (Interestingly, if the folk village is supposed to reflect late-nineteenth-century Korea, there is no compelling reason for it not to include a Christian church.) Ultimately, the discourse was one of *minjungjuŭi,* at times tempered with a dose of North Korean *juche*—the idea of self-sufficiency espoused by Kim Il-sŭng and his son Kim Jong-il—and not the discourse of *munhwa minjokjuŭi* that characterized my walks with Kim Min Sun. Of course, unlike the North Korean parks, the best known of which is Mangyŏngdae, where the focus is on the leader and his humble beginnings (a park that finds its South Korean counterpart in the birthplace of Pak Chung-hŭi), this reading of the folk village espoused no such political elitism and focused quite deliberately instead on the *pot'ong saram*—the everyman.

Because of the success of the prodemocracy movements of the 1980s, there were profound changes in the Korean political and cultural landscape. Among the many changes attendant to this burgeoning democracy were significant reforms in labor practices and a concomitant increase in leisure time available to the ever-growing middle class (Ministry of Transportation 1994). The 1987 presidential elections also ushered in a substantial loosening of government control in the economic, political, and social arenas, a loosening that has continued in the years since. The 1988 Seoul Olympics was touted as the opening of Seoul to the world and was grafted onto the emerging discourse of globalization—*segyehwa.* The Olympics also saw the first significant influx of foreign tourists in modern Korean history, an influx reinforced by the 2002 World Cup. Along with these changes in the cultural landscape came a shift in the location of Minsokch'on in the public imaginary perhaps best expressed in the spatial relationship of the folk village to the nearby megatheme park and its transformation from "Yongin Farmland" to Everland. In recent years, rather than being primarily an institution that imbues domestic visitors with a sense of national pride, Minsokch'on has also become more and more of a theme park—a place for modern people compliant with the park's main narrative to examine the quaint customs of a long-forgotten past and, at the end of the day, to buy a small piece of the past to bring home and hang on the apartment wall of the present.

The theme park, the folk village, and this form of global consumerism are drawn together under one roof at Lotte World in Chamsil, described in a recent tourist brochure as "a mammoth indoor leisure and shopping mall . . . composed of a . . . folklore center, indoor theme park . . . shopping mall and department stores." Built by a Korean conglomerate as part of the massive development south of the Han River in the lead-up to the 1988 Seoul Olympics, Lotte World has become a mainstay of both tour groups and the adolescent mall culture that has grown up in contemporary Seoul.

Perched atop the mall's glass galleria, with its always crowded skating rink, its many chain stores, and its unusual diversions, such as a shooting range, sits a re-creation of a crazed fantasy of Holland, Baghdad, and the Bavarian Alps named Adventureland; across a footbridge on a man-made island is another amusement park: Magic Island. Fittingly, the main feature of the island is a castle that sits halfway between Mad King Ludwig's Neuschwanstein and Disney's Sleeping Beauty Castle. The itinerary through the bottom of the complex takes one up through the colder reaches of the skating arena, the stores of the second and third levels, the impossible balloon ride through the windmills of the amusement park, and the monorail out to the castle. At the very top of the complex is the more staid Folklore Museum that provides, in miniature, a glimpse into the "folk culture" of Korea. The spatial organization of this folklore museum is one that resists anything but absolute compliance—this does not mean that one cannot engage the space playfully, but unless one is willing to climb large barriers, there is no escape from the narrative presentation of Korea that takes one from the Cretaceous period, where we learn that the Korean Tyrannosaurus was three times the size of the North American variant (a narrative that stretches far beyond the normal 5,000 years of Korean history and also ignores the fiction of the Tyrannosaurus), through a Stone Age dwelling, the Bronze Age, and a series of rooms dedicated to the earliest kingdoms of Korea. The itinerary then takes one on to the spectacular "miniature village." Despite the size, these small, almost caricature-like models of people and folk customs might be more honest than the painstaking reproductions at the folk village. Here at Lotte World, folk culture is nothing more than a diversion, small, packageable, and saleable as part of a narrative of Koreanness that concatenates consumption, tourism, and a sense of "something unique" that can be understood—and enjoyed—as a spectacle, far removed from the "real life" of the shopping mall, ice rink, and amusement park downstairs.

The tentacles of this contemporary representative endeavor that keys on "Chosŏnness" as part of the commodification, at times even fetishization, of the past, reach out in all directions in modern Korea and exert their grip in areas as diverse as the media (through film, television, and advertising), the city (in the guise of cafés and restaurants), and popular culture (through video games, pop music, and the covers of—oddly enough—punk rock CDs). Appadurai, describing this reach of the past into these modern realms, opines that "the past is now not a land to return to in a simple politics of memory. It has become a synchronic warehouse of cultural scenarios, a kind of temporal central casting, to which recourse can be taken as appropriate, depending on the movie to be made, the scene to be enacted, the hostages to be rescued" (1996, 30). The tentacles of this synchronic warehouse reach even further than the very public urban landscapes and mediascapes, grabbing hold of illegal economies such as room salons, gambling rooms, and brothels that in various guises play on received notions and projected ideas about this rural late Chosŏn idyll.[12]

Minsokch'on, the folk museums, Everland, Lotte World, and the myriad yet less developed representations of folklife that abound in contemporary Korea ultimately stand as remarkably complex loci intimately linked to the cultural politics of display, accessed by spectators in various ways as resources for the negotiation of individual identity, implicated in the competing nationalisms that characterize Korean ideological debates, and bending in the fierce winds of globalization. With the rapid changes in the political and cultural landscape over the past ten years, individual spectatorship and subsequent interpretations at sites such as the folk village have necessarily changed. Ultimately, one cannot propose a single position of spectatorship at any of these sites—rather, one must recognize that visitors move between the complicit and the resistant, the compliant and the ludic. The history of the development of the field of folklore in Korea, the politics attendant on the earliest display of Korean rural life for non-Korean audiences, the role of the earliest Korean folklorists as cultural nationalists fighting Japanese colonialism and the policies of cultural erasure, the competing claims to the discourse of Koreanness in the postcolonial era, the division of the country by warring superpowers, the later emergence of the folk museum as an institution intended first for domestic edification and then foreign education, the more recent incorporation of folklore and folklife into the open-air museum, and the ever-closer connection between the folk village and the theme park all play important roles in the insertion of Korean rural folklife into the competing definitions of what it means to be Korean. Subsequent readings of these sites by scholars, by Koreans from all walks of life, by international tourists, and by immigrant workers, as well as reinterpretations—at times collaborative, at times resistant—of these cultural displays and the translation of displays of rural life into locations as diverse as shopping mall attractions, cafés and restaurants, television dramas and computer games, bars and room salons all reveal the extent to which conceptions of the Korean "folk" are central in understanding the complexities of contemporary Korean society. Folklore and folklife are not peripheral phenomena in Korea, as they might often appear to be in other countries. Rather, the control and manipulation of the symbols attached to Korean folklife are hotly contested battlegrounds between rival institutions, between generations, between classes, between genders, between individuals, and even between countries.

Notes

I would like to thank members of the Wildcat Canyon Advanced Seminars in Cultural Studies for comments on earlier drafts of this essay.

1. The village was established in May of 1973 and opened in October of 1974. Yi Yong-sop (1994) provides an excellent, albeit outdated, overview of Korean museums.

2. Bjarne Stoklund (1993) contests the importance of Hazelius in the emergence of the folk museum (see also Sandberg 2003).

3. The updated Korean National Folklore Museum from 1993 has shifted some of these displays to include technological innovation (Kendall 1999; Kungnip Minsok Pangmulgwan 1994).

4. The "will to authenticity" aligns with MacCannell's discussion of authenticity in tourist experience (1976). Ning (1999) proposes a more complex understanding of a range of authenticities in various forms of the tourist experience.

5. Hendry (2000, 134) notes a similar genealogy for certain Japanese parks, a genealogy that has also reached to other Asian countries (Hitchcock 1995).

6. National Treasure number one is Namdaemun, so designated on December 20, 1962.

7. Peasant revolts in 1812 and 1862, the aborted reform attempts of 1884, crop failures of 1889 and 1891, the Tonghak rebellion of 1894, increased foreign pressure on Korea in trade, a crumbling monarchy, and increased Japanese incursions into the region are just a few of the events that countermand the image of the nineteenth century as an idyllic period in Korean history.

8. I was unable to find any information about Yi Tok-kyu in the standard

biographical dictionaries such as the *Kuksa taesajŏn* or the *Han'guk inmyŏng sajŏn.* I would like to thank Professor John Duncan for his assistance in trying to locate information about Yi.

9. Hendry's (2000, 151–178) discussion of similar phenomena at Japanese museums resonates with the Korean situation.

10. Cohen (1979, 183) provides a similar typology of tourist experiences, positing five experiential modes: (1) the recreational, (2) the diversionary, (3) the experiential, (4) the experimental, and (5) the existential. In Cohen's model, the differentiation of standpoints is somewhat linear, and there is little appreciation of the movement between interpretive registers that characterizes individuals' engagements with the folk village, both over repeated visits and during a single visit.

11. See also Kendall 1996.

12. It is not difficult to find, for example, room salons and brothels catering to a primarily Korean clientele (as opposed to primarily foreign clientele) concentrated in areas of Seoul such as Miari, where young women and girls are forced to wear *hanbok* and sing *minnyo* (folk songs) to their customers, in many cases proposing to create an environment reminiscent of a village inn amid the urban sprawl of Seoul.

5

Blame Walt Rostow

The Sacrifice of South Korea's Natural Villages

DAVID J. NEMETH

We have fought against nature stubbornly and managed to conquer it at last.

—Kim Young-soon

THE CONCEPT OF a "natural village" appears in Korean, Chinese, and Japanese development literature and is officially a data collection category in many of their statistical yearbooks. The *Cheju Statistical Yearbook* (Cheju Provincial Bureau of Statistics 1985), for example, uses the term "natural village" *(chayon purak)* to describe all villages identified as administrative units on that island province. However, throughout Northeast Asia, as on Cheju Island, a "natural village" is informally understood to be a primal settlement, or hamlet, now extinct: an isolated self-governing rural settlement with strong local identity, comprised of inhabitants whose intimate community was characterized by constant and face-to-face social interaction, wherein a village deity or common ancestor provided a strong centripetal force for their community and repetition of group ritual reproduced and sustained local isolation and self-sufficiency. They were natural communities, though human centered in an ecological sense. Indeed, extinction of the natural village in South Korea can be dated exactly to the year 1976, and it continues to be celebrated as a human scientific achievement, part of a story of a "miracle" and thus somewhat akin to the eradication of smallpox; with a truth so seemingly self-evident, it is beyond reproach.[1]

The Neo-Confucian Natural Village

Anyone visiting the hundreds of villages in Cheju Province today might wonder what, if anything, is "natural" about them. This is because the structures in the contemporary built environment appear so new and the inhabitants seem so cosmopolitan. As recently as fifty years ago, on Cheju Island and throughout much of South Korea, the term "natural village" had a different and tangible meaning. At that time a natural village retained some of the unmistakable neo-Confucian stamp of its premodern organic unity. There were still traces of the neo-Confucian past everywhere in the salient surroundings—in the sights, smells, and sounds. Experiencing these tangibles of local identity, steeped in neo-Confucian symbolism, was then still possible. A visitor could still sense the power and spirit of the neo-Confucian planning model that long ago constructed and shaped the rural hamlets. No more; now that the natural villages are extinct, that experience is gone and irretrievable.

The difference between a classical South Korean neo-Confucian natural village and a modern South Korean administrative unit called a "natural village" is therefore profound. Their simultaneous incompatibility in space and place exemplifies the theme of contested spaces and constructed places addressed by the present volume's theme of critical approaches to Korean geographies that gave shape and direction to this research. I begin my presentation by briefly acknowledging the successes and sacrifices of South Korean activists participating in Minjung Undong (the People's Movement) protests and critiques during the 1970s and 1980s.[2] My firsthand observations of indigenous forms and multiple targets of popular critique during those years served to direct my attention to the plight of South Korea's natural villages and their inhabitants. The vivid example of Minjung Undong critique has inspired my outsider's critical commentary, which is presented here.

Minjung Misul as Critique

Minjung Undong, as I observed it in action in on Cheju Island, was broad in scope and diverse in composition and approaches. Literary, artistic, and theatrical activities were together called Minjung Misul Undong (the People's Art Movement). Block print creations from hand-carved wood and linoleum stamps designed for clandestine mass production and distribution, Minjung Misul critiques were designed to speak for themselves, and they did so very often with shocking clarity, revealing the dark underbelly of the Korean modernization story.

For the purposes of the discussion here, South Korean modernization began shortly after the advent of modernization theory and with the implementation of the first five-year plan in 1961, under the Park administration. Rural modernization began in earnest with the heavy-handed Saemaŭl Undong (New Village Movement) in 1971. Collateral damage thereafter was fast and fierce, resulting in the emergence of the Minjung Undong and its various forms of antigovernment critique and protest, including Minjung Misul Undong (for an overview of the Minjung Undong, see Abelmann 1996). I purchased several Minjung Misul art block prints from activist students in 1979 and again in 1985. More recently, a large compendium of diverse examples of Minjung Misul protest art has been compiled by Kim Tae-kyong and privately published (1988). The typical block print critiques modernization, depicted for example in one print as a monstrous, berserk bird, part natural and part machine, seemingly intent on destroying its own young.

Neo-Confucian Space

East Asian agricultural production systems are notable in human history for their political stability, for their ability to support large human populations through intensive agricultural practices, and, according to the geographer Joseph Spencer (1954, 315), for having provided these populations with the highest quality of life for human populations up to the eighteenth century. Francesca Bray concurs: "Only after 1800 did rural living standards begin sharply declining" in China (1994, 36). South Korean neo-Confucianism derived from a twelfth-century Chinese synthesis of Taoist, Buddhist, and Confucian ideologies and is widely considered to have been the most orthodox and enduring example of neo-Confucianism in East Asia. I have often argued and attempted to demonstrate that "enlightened underdevelopment" was a successful guiding principle of neo-Confucian agricultural civilizations and that it played a major role in creating a natural village landscape that could be described in neo-Confucian perspective as a *song*, or "sincere" landscape populated by *in* or "virtuous" people preoccupied with cultivating *yeui*, or "propriety" (Nemeth 1987, 171).

In brief, cultivating propriety through enlightened underdevelopment thinking involves accepting the principle and practice of deliberately rejecting labor-saving devices as a demonstration of one's virtue. Enlightened underdevelopment thinking presents a direct challenge to economic growth ideology to the extent that it provides a critique of the pursuit of those "wants" disguised as "needs" that drive change. Traditional standards of virtue related to enlightened underdevelopment thinking can perhaps be reinstated in Asia through building popular consensus against developmental ideology. For example, the continuing Japanese economic slump is now being interpreted as a critique of the last century and a half of rapid change that has undermined Japanese stability and cultural vitality, leaving a strong nostalgia nowadays for the Edo Period, which is popularly perceived as a time when life in Japan was "more enjoyable" (Onishi 2004). The assumption underlying enlightened underdevelopment thinking is that a person is born selfish and must be educated in the classical canon to work for the good of the community. To illustrate, consider the example of the following "Story of the Swape."

> Tzu Kung had been wandering in the south in Chuu, and was returning to Chin. As he passed a place north of the Han (river), he saw an old man working in a garden. Having dug his channels, he kept on going down into a well, and returning with water in a large jar. This caused him much expenditure of strength for very small results. Tzu-Kung said to his, "There is a contrivance . . . by means of which a hundred plots of ground may be irrigated in one day. Little effort will thus accomplish much. Would you, Sir, not like to try it?" The farmer looked up at him and said, "How does it work?" Tzu-kung said, "It is a lever made of wood, heavy behind and light in front. It raises water quickly so that it comes flowing into the ditch, gurgling in a steady foaming stream. Its name is the swape." . . . The farmer's face suddenly changed and he laughed, "I have heard from my master," he said, "that those who have cunning devices use cunning in their affairs, and that those who use cunning in their affairs have cunning hearts. Such cunning means the loss of pure simplicity. Such a loss leads to restlessness of the spirit, and with such men the Tao will not dwell. I knew all about (the swape), but I would be ashamed to use it" (Needham and Wang 1965, 332–334).

The passage originates in the Taoist classic *Chung Tzu* and dates from about the fifth century BCE. The swape in this story is a counterbalanced bailing bucket, which was gradually and successfully introduced into China as a labor-saving device for subsistence farmers.

The "walking tractor" (a mechanical hand tractor or rototiller) is in essence, if not function, a modern-day swape—a multifunctional labor-saving device that has played a dramatic role in promoting economic development throughout Asia and in South Korea.[3] During the late 1950s, Dr. Forrest R. Pitts, now considered the "father of the hand tractor in Korea" (Pitts 2002, 282–283) was reconnoitering the countryside to establish some strong arguments in favor of introducing rototillers into South Korea: "We encountered this young farmer, and I told him that I was interested in how Korean farming could be mechanized, via the rototiller. He told us with considerable energy, all the while being calm about it, that the Korean red cattle were an integral part of Korean farming, and that the animals were considered almost part of the family. He said he hoped that I would reconsider my ideas. I can't remember where that was" (Pitts, pers. interview 2004). Hand tractors began to be imported in 1961, in advance of the founding of the Saemaŭl Undong. The swape was eventually widely adopted by farmers throughout Northeast Asia. Widespread adoption of the swape should not, however, imply any widespread rejection of the ethical warning. The enlightened underdevelopment concept carries with it the principle of rejection but is not antitechnology. Change is inevitable, which is why the *Book of Changes (I Ching)* is the principal classic of the neo-Confucian canon.

Neo-Confucian ideology is infused with the ethics of enlightened underdevelopment mainly through Taoist spiritual influences, which often act contrary to the selfish impulses of Confucian pragmatism. The Taoist principle of rejecting labor-saving devices for ethical reasons creates an antitechnology paradox in neo-Confucian agricultural civilizations where efficiency is also praised and encouraged. The gradual process of the popular acceptance of the swape, in the context of a Taoist cosmology, might be described as a gradual process of "anthropomorphism" (an inanimate object—the swape—over time becomes socially constructed as animate) in contrast to a process of "counteranthropomorphism" (where an animate organism over time becomes socially constructed as an inanimate object; for example, through the process of sacrifice). This process of transformation is introduced as "situational definition" and further discussed below.

P'ungsu and Enlightened Underdevelopment Thinking

While difficult to articulate in words, cosmological diagrams, *p'ungsu* (Chinese *feng-shui*) maps, and humanized landscape paintings by traditional East Asian artists have often captured in their idealized images and arrangements the perfect outcomes of prolonged enlightened underdevelopment thinking. What is striking about the neo-Confucian natural village and is revealed in East Asian landscape paintings is the achievement of human productivity in splendid isolation. This natural village in turn is seen as the ideal subsistence society, where "everything produced locally is consumed locally." The "sincere" neo-Confucian landscape should be seen as an ideological construction. Artifacts of the "sincere" cultural landscape were "architectures of ideology" constructed as didactic (teaching) and mnemonic (memory) devices that influenced farmers as they negotiated the landscape. They reproduced the dominance of neo-Confucian ideology, in cyclic fashion with the seasons, year upon year. *P'ungsu* practices played a major role in this reproduction process.

Despite this long historical construction of the "sincere" landscape, modernization theory systematically destroyed it, through the sacrifice of the neo-Confucian natural villages. Although my critique is directed at developmental ideology in South Korea for its violence against natural village manifestations constructed by neo-Confucian ideology, Korean neo-Confucian ideologues were in their heyday no less harsh in wresting space from shamanism and Buddhism in order to construct neo-Confucian space.

On Cheju Island, for example, there was a violent neo-Confucian purge of shaman shrines and Buddhist temples during the early seventeenth century. To construct a place, you have to clear a foundation.

Theorizing Sacrifice

The sociologist Arnold B. Arluke proposes that sacrifice "involves the transformation of the victim into both object and myth. . . . The victim goes through an objectifying metamorphosis culminating in its death and contribution to the larger community. It assumes a new form albeit no longer corporeal. This metamorphosis entails the stripping away of the victim's former nature, such that the purest body remains and can be used in a generalized manner. . . . [L]iving beings become objects. At the same time, the victim must be linked back to the community in symbolic form" (1988, 99). Arluke devised this particular theory of sacrifice to describe and explain the process by which laboratory animals were objectified during preparation for scientific experiments. He describes their objectification as an exercise in "counteranthropomorphism," a process of psychological restructuring that prepares the path for the physical restructuring that is sacrifice. Once the living organism is successfully objectified, sacrifice is possible and rarely opposed. This objectification process has five parts: incorporation, deindividualization, commodification, isolation, and situational definition.

Although Arluke theorized sacrifice to explain how medical lab technicians both rationalized and ritualized their day-to-day activities with animals, the same approach has clear application to the process of developmental planning wherever natural villages were involved. That is, *if* natural villages had organic identity, then some of Arluke's observations might well apply. Arluke's theory of sacrifice provides, at the very least, an unusual way to critically map the fate of South Korea's natural villages. I briefly consider here Arluke's five-part psychological preparation process of counteranthopomorphism as it might apply to natural village sacrifice:

1. *Incorporation.* Incorporation of natural villages into the South Korean modernization plan involves measuring them for future economic planning purposes. As their identity becomes redefined from a "living organism" to an "administrative unit" comprised of the sum total of their planning data, they are stripped of the mystical power of their organic identity. They are rendered into no more than the sum of their parts. An artificially imposed pseudospirit replaces their innate spiritual quality.
2. *Deindividualization.* Reassignment as administrative villages effectively deindividualizes natural village organisms in place. Deindividualization redefines the significance of their existence. It facilitates their

objectification by forcing them to be merged into classes, where they are no longer referred to as unique organisms and where they can be discussed and manipulated as abstract entities.

3. *Commodification.* Commodification turns natural villages into secular objects to be freely exploited. Natural villages first are defined as expendable objects by developmental planners and then perceived as such by their inhabitants. Commodification denies the organic structure of the natural village and its inhabitants through the process of secularization. Villagers that most resist objectification are targeted as those most in need of psychological and physical rehabilitation. Natural villages that persist to resist modernization in place may be forcibly evacuated and reconstructed on new sites. The reason for this draconian measure is that the neo-Confucian natural village as a social construct was literally a natural community of blood lineage grounded in ancestral land (Chung 1988, 45): the three Sino-Korean characters describing that society translate as "spirit," "soil," and "sky" and indicate "a people gathered round on the earth under which lies the spirit of the ancestor." Contesting neo-Confucian space by moving local ancestral tombs can also break the spiritual bond between the farmer and the natural village and speed up the developmental agenda.
4. *Isolation.* Isolation involves the separation of a few natural villages from other rural settlements for experimental purposes. These are called "model villages" by developmental planners—examples of excellence to be emulated by other villages. They have the honor of being the first to be sacrificed. In addition to the isolation of a few natural villages as model villages, a few are also isolated as "folk villages."[4]
5. *Situational definition.* Situational definition of a natural village as a developmental administrative unit is accomplished by promoting zealous psychological defense of the developmental point of view in place, ensuring its entrenchment. Constant meetings are held to educate and praise change and to disparage old thinking. Patriotism for the nation-state is cultivated. Instrumental-rational thinking is rewarded. Unenthusiastic farmers become ostracized and discriminated against.

These then are the five procedures that accomplish the goal of counter-anthropomorphism, a process that denies the organic identity of the natural village in place by systematically alienating the farmer from his root sources of sustenance and inspiration in the ancestral soil. Moving from Arluke's abstract model of sacrifice to some relevant empirical research, a most pertinent example is provided by Jeong-man Lee (1991), who has described in detail the sacrifice experienced by two generations of South Korean farmers. His case study of Wonpyongchon Village in mainland South Korea is titled "The Generation of Sacrifice: Modernization and Korean Farmers." Having here linked the sacrifice concept to modernization, it is appropriate at this juncture to turn to the concept of "modernization" and to the workings of modernization theory.

Contesting Neo-Confucian Space in South Korea

"Modernization" is one word that describes the complex process of deliberate change that promotes economic growth through industrialization. South Korean economic analysts often associate modernization with another term: "developmental ideology." "Economic growth ideology" is also a useful phrase that describes the powerful idea driving the process of modernization, which in the South Korean example contests not only space but time.

Modernization *is* developmental, and it is a product of Enlightenment thinking in European intellectual history. It gradually evolved into a comprehensive worldview—a Western Enlightenment cosmology—characterized by distinctive philosophical and political assumptions and ambitious goals. Time in the model of Western Enlightenment cosmology is perceived as linear in contrast to cyclic, and this distinction is perhaps the most profound difference worth noting when describing the contesting of neo-Confucian space by development ideology in South Korea between 1961 and 1977.

Arguably, modernization theory or development ideology was first introduced in 1960 by Walt Whitman Rostow (1916–2003), in his book *The Stages of Economic Growth: A Non-Communist Manifesto* (Rostow 1961/1990; see also Purdum 2003). The book promoted a clearly articulated planning model for replacing traditional agricultural landscapes, structures, and psychologies with industrial landscapes, structures, and psychologies. The following list describes the highlights of Walt Rostow's five stages of economic growth, revealing it as a product of developmental ideology.

Stage 1: Traditional Society
- Most workers are in agriculture;
- Subsistence economy;
- They have no savings;
- They use old techniques and technologies (tools).

Stage 2: Preconditions for economic "takeoff"
- Awakening the farmer to aspire to a "better life;"
- Increased production (leading to surpluses and savings);
- Individual values and attitudes change from "I-Thou" to "I-it";

National goals are changed to privilege manufacturing over agriculture.

Stage 3: Economic "takeoff"

New technologies are applied to further increase production;

Enthusiastic support for manufacturing industries; Rural to urban migration increases;

Dramatic improvements in the infrastructure (transportation, education, communications);

Political power is transferred from a landed aristocracy to manufacturing elites.

Stage 4: Drive to economic maturity

Urbanization continues;

Service industries grow;

Rural areas are depopulated; rural areas are highly mechanized.

Stage 5: High mass consumption

Emphasis shifts to non-economic pursuits, recreation;

Production is eclipsed by consumption.

Rostow's five stages of economic growth, as a product of developmental ideology, reveal "development" to be an unsubtle path for national economic development. In his model, planners use their almost unlimited power to achieve changes that move toward the kind of social and economic system that their country decides "it needs" (Schramm and Lerner 1976). Since "needs" and "wants" are constantly created by the capitalist system in order to increase profits, development that follows Rostow's five stages is a clearly articulated means in search of poorly articulated ends. The remainder of this chapter focuses attention on Rostow's stepwise modernization theory as historically applied by South Korean economic planners, resulting in the intentional eradication of subsistence farming, a bedrock of the "natural" village and "sincere" landscape.[5]

Saemaŭl Undong

In South Korea, the spearhead of industrialization ideology and antitraditionalism in the rural sector of the economy has been the government's Saemaŭl Undong or New Village Movement, launched in 1971. Keim describes the ideological thrust of Saemaŭl Undong: "to advance national solidarity by eliminating the cultural imbalances between the urban and rural communities" (1979, 19). The government perceived that while South Korean cities were "Moving ahead at a dizzying pace . . . the countryside was still sunk in lethargy, passivity and even cynicism. Rural people were straggling behind their urban brethren. Something had to be done to correct the situation. The farmer had to be awakened" (Kim Youngkwom and Lee 1979, 583). This "awakening" was a harsh one and entailed prodding neo-Confucian farmers, however reluctant, into becoming Western instrumental-rational thinkers in order to increase their productive potential. "Authoritarianism reigned in Saemaŭl projects," according to John Lie (1998, 110). Initial success in the rural sector, in large part due to this forced compliance, later inspired lofty goals for expanding Saemaŭl Undong into the cities. However, by the early 1980s the movement had become so bogged down in a morass of corruption that it was disbanded (Chung and Veeck 1999, 276).

If Rostow's five stages were intended as an economic planning model for South Korea and other Third World countries during the Cold War, this forced awakening of traditional farmers during Stage Two seems nothing less than a planned act of aggression against traditional farming cultures on a worldwide scale. Unlike the disruptions of typhoons, insect invasions, and volcanic eruptions, the likes of which traditional Korean farmers could comprehend through many centuries of group experience as natural disasters related to fate, luck, *p'ungsu,* and virtue, Saemaŭl Undong violence was unprecedented, unnatural, and premeditated—and incomprehensible to farmers, whose orderly lives were suddenly disrupted and sent spinning into a downward spiral. Occasionally, critics of modernization, when pointing to collateral damage deliberately inflicted on traditional farming cultures by economic development planners, call it a planned genocide. A close reading of the Genocide Convention of 1948, especially as it is now interpreted, validates their critique.[6]

Chronology of Natural Village Sacrifice

The sequence of Saemaŭl construction rites and stages of promotion that constitute the planned process of natural village sacrifice was linear (Table 5.1). Close examination of this sequence reveals how modernization ideology successfully contests neo-Confucian ideology in South Korean space and constructs South Korean places, in part, by using old words in new ways. Indeed, clever word substitution in this table effectively strips "natural village" of its clarity of meaning in the neo-Confucian cosmological context and thus voids it of its didactic and mnemonic power to reproduce neo-Confucianism in place. By implying that traditional natural villages originating in neo-Confucian context are "underdeveloped," in contrast to "developing" or "developed," South Korean economic planners deny their neo-Confucian identity as mature and healthy organisms and thus rob them and their inhabitants of their dignity.

For example, village promotion in the category of "roof improvement" requires the substitution of tin or plastic roofs for grass roofs. Yet grass roofs once affirmed in the eyes of virtuous neo-Confucian farmers that the landscape they cultivated so intensely was sincere. The grasses, produced as byproducts of local grain production, were consumed locally not only for roofing materials but in the local manufacture of raincoats and shoes and for mixing with animal dung to make compost to be used as organic fertilizer. Mark Clifford describes the "end of the thatched roof era" and directly attributes its demise to aggressive, myopic Saemaŭl strategy and its implementation, and he specifies some of its negative impacts on rural quality of life:

> Saemaŭl was a product of the military reformist spirit that animated [President Park] and the men around him. Away went the thatched roof huts that had dotted the countryside for centuries. Park considered these roofs too primitive for a would-be developed country, so he ordered their replacement with corrugated metal roofs and, later, concrete tiles. The fact that straw was a far more effective insulator against the bitter winters and steaming summers of Korea did not trouble Park. Metal was modern, and straw was not. For those villagers who protested, Park offered a sop: the Korean Folk Village . . . where traditional houses were preserved for future generations. Peasants who refused to re-roof their houses had their traditional roofs destroyed by zealous local officials (Clifford 1998, 94–95).

TABLE 5.1. Natural village sacrifice: Sequence of construction rites and stages of promotion

Project	**Promotion from underdeveloped village to developing village**	**Promotion from developing village to developed village**
Village road	Completion of main road	Completion of main and feeder roads
Farm road	Feeder farm road into the village	Main farm road to the village
Bridges	Small bridges (10 m+)	Small bridges (20 m+)
River control	Small streams passing through the village	Small rivers near the village
Community facilities	Village hall, storehouse, workshop, etc.: 1–2	Village hall, storehouse, house, workshop, etc.: 2+
Roof improvement	50% of all houses	80% of all houses
Saemaŭl Fund	Over 300,000 won	Over 500,000 won
Savings in Saemaŭl Fund per household	Over 10,000 won	Over 20,000 won
Annual income per household	Over 700,000 won	Over 900,000 won

Source: Revised after Ministry of Home Affairs (1980: 2, 23–24); Whang (1981, 183).

The last South Korean natural village ("underdeveloped village") was sacrificed sometime during 1976 (Table 5.2).

During the height of the Saemaŭl movement, that grass suddenly became a disposal problem instead of a resource for developing villages. At that time, frequent bonfires in the fields created smoke so thick in the villages that inhabitants could hardly make out the brightly colored new plastic roofs. Boosters of Saemaŭl progress were deliberately blind to its scorched-earth policy, just one indication of an eradication of an older way of life. Instead, Saemaŭl Undong has been persistently acclaimed over the decades by almost all South Korean scholars—and most North American scholars—for contributing to an "economic miracle" and dramatic benefits the rural sector rendered in landscapes of plastic and cement (Turner et al., 1993, 84–85).

There were and are few incentives for scholars to listen more closely to the stories of sacrifice told by those many individual farmers who were forced out of their rural mindsets and enclaves, often in despair and desperation, to survive on low-wage manufacturing jobs in the cities. As with other past atrocities that are now being openly discussed thanks to critical approaches in the social sciences, the atrocities related to the Saemaŭl Undong may eventually be told and become a topic for public debate.

Blame Walt Rostow

One of the most effective techniques of the Minjung Misul critique, and one that was not at all neo-Confucian, was that of direct accusation (Fig. 5.1). This rude but effective finger-pointing can apply here too, as one can point an accusatory finger at Walt Rostow for the sacrifice of South Korea's natural villages. Indirect evidence in support of my own finger-pointing is pro-

TABLE 5.2. Tracking the pace of natural village sacrifice

Year	Total	Underdeveloped village	Developing village	Developed village
1972	34,665	18,415	13,943	2,307
1973	34,665	10,656	19,763	4,246
1974	34,665	6,165	21,500	7,000
1975	35,031	4,046	20,936	10,049
1976	35,031	302	19,049	15,680
1977	35,031		11,709	23,322
1978	34,815		6,114	28,701
1979	34,871		976	33,893

Source: Revised after Ministry of Home Affairs (1980: 2, 23–24); Whang (1981, 185).

vided by comparing his five-stage planning model to the Saemaŭl Undong method of operation, and strong evidence that links Rostow directly to the genesis of the Saemaŭl movement is in his last book (2003), one chapter of which is entitled "My Marginal Association with a Miracle." In this chapter, the extent of his association with economic growth ideology and its implementation in the Republic of Korea becomes clear; the impact of his ideas appears to have been far from "marginal." The central role of Rostow's Stages in that conspiracy of economic growth now hailed as "The Korean Miracle" is obvious in a speech that he delivered to the students and faculty of Seoul National University on May 3, 1965, shortly after his private audience with the architect of Saemaŭl Undong, South Korean president Park Chung-hee. President Park and his technocratic young bureaucrats, as history bears out, were not sentimentalists with nostalgia for the past—except where the past might fit into a museum setting. As Rostow relates, "Park was interested in my SNU Speech—how would I deal with the question of Korea's place in the *Stages?* I responded that South Korea, in my view had begun its takeoff. . . . It was my conclusion as an American planner that South Korea was in the early phase of takeoff, and it could look forward to a high rate of growth. The speech attracted an overflow crowd, and drew

FIG. 5.1. Artist and title unknown (Kim Tae-kyong 1988).

media attention. My conclusion about where South Korea stood in terms of the stages of growth was picked up by the Korean press" (2003, 257–258).

The people of South Korea could rest assured in Rostow's testimony that they were by then far beyond Stage 1, that rapidly fading mystic realm of intentionally distorted memories to which there would be—could be—no return.[7]

Notes

1. Cheju Island is located in the northern reaches of the East China Sea, between China and Japan, and approximately 60 miles off the southern coast of the Korean Peninsula. I lived and worked there from 1973 to 1975, from 1979 to 1980, and again from 1984 to 1985. My tenure on Cheju overlapped with an approximate decade of significant socioeconomic change on Cheju Island that I still consider remarkable and troubling, as I bore witness to rapid and ruthless transformations of Cheju's traditional natural villages into modern administrative villages. My general impressions of qualitative gains and losses during this period of transition, with emphases on the processes, outcomes, and implications of ideologically inspired anthropogenic changes, are presented in this chapter.

2. Antigovernment protests organized by farmers during these decades became increasingly direct and violent as farmers gradually became aware of their impotency against relentless Saemaŭl advances, driving them to desperate measures (*Korea Times* 1989).

3. The story of their insidious impact on subsistence agricultural systems as a "Trojan horse" for the Saemaŭl invasion of the natural village is related in Nemeth (1988).

4. Tangherlini's chapter in this book focuses on Korean folk villages, so I won't elaborate on that topic here (Tangherlini this volume).

5. Other examples of the use of Rostow's Stages as a planning model exist throughout Asia and elsewhere. Contemporary events in India demonstrate that the same stepwise process of creative destruction is now taking place there (Waldman 2004).

6. Adopted by Resolution 260 (III) A of the United Nations General Assembly on 9 December 1948. Article 2c, for example, would seem to apply to traditional farmers as a targeted group, to the extent that the South Korean government by following Rostow's five-stage economic planning model was guilty of "deliberately inflicting on the group conditions of life calculated to bring about its physical destruction in whole or in part." This would be a "punishable act" according to Article 3b because there was a clear "conspiracy to commit genocide" (United Nations General Assembly 1948).

7. "No return" to Stage 1 is implicit in Rostow's manifesto. This was apparently lost on Dr. Zo Zayong of South Korea, today a famous tragic figure, who dedicated his life to saving the "folk art, customs and culture" that had been sentenced to

death by Saemaŭl Undong edicts. Dr. Zo "was sternly opposed and even threatened" by the South Korean government during his activist days. He was a strong advocate of Minsok Munhwa (the People's Cultural Heritage Preservation Movement), and in fact founded an "Old Village Movement" as a grass-roots alternative to the government's New Village Movement. He was an educator who "tried to educate young Koreans about their ancient traditions and inspire them to respect them" (Mason 1999b).

Part 3

Geographies of Religion

6

Auspicious Places in a Mobile Landscape

Of Shamans, Shrines, and Dreams

LAUREL KENDALL

THE FOLLOWING STORY falls somewhere between a field anecdote and a fairy tale.[1] As anecdote, it is reconstructed from field notes, transcripts, and memory without conscious embroidery, elaboration, or fabrication. As a fairy tale it resembles a genre of stories sometimes attributed to Buddhists or Taoists, where illusions are at play and a lesson may be learned by confronting them. Such tales were very much with me as these events unfolded.

An Old Shaman in a Tile-roofed House

In the summer of 1994, I was making the rounds of commercial shaman shrines *(kuttang)* in the mountains surrounding the city of Seoul, attempting to survey the changes in this world since my first fieldwork in the 1970s. Kim Sung Ja, my field assistant in these endeavors, is an amateur folklore buff with a keen interest in the shaman world and in old Seoul traditions. When a blind diviner mentioned a venerable shrine, we were both keen to find the place. We emerged from the subway station in an urban neighborhood that had been a rural village not so long ago. We asked directions to "a *kuttang,* a place where shamans hold *kut.*" Some people gave us only blank stares, while others nodded with recognition, but their directions sent us down blind alleyways. The shrine seemed to retreat from us, like a mirage. Finally, in the district office, a gaggle of old men affirmed that there has been a shrine in the neighborhood for three or four centuries. A scholar built it to memorialize his daughter who had died before her wedding ("built it to placate a mischievous virgin ghost," I thought). In the past, the old men had all participated in the annual community ritual for the tutelary god in the shrine, but the neighborhood no longer sponsors the ritual and the shaman has been left to honor the tutelary on her own. The old men indicated a hillside path on the far side of a broad boulevard and told us that the shrine was just below the old people's home on the summit.

We followed the winding lane through a neighborhood of modest houses, none of them more than twenty years old. Just below the summit, we came upon a crumbling tile-roofed house set apart in some tall grass and surrounded by a wall—a house from another time. The big wooden gate was marked with the *taeguk* symbol, the circular swirl of red, blue, and yellow that signifies the cosmos and marks a shrine. This must be it. We knocked and shouted, but the place seemed to be deserted. Ms. Kim asked me if I had ever read Kim Tong-ni's novella *Portrait of a Shaman*. The evocative description of a shaman's house had been running through my head as well: "an antiquated tile-roofed house with one of the upper corners already crushed out of shape. On the roof tiles mushrooms sprouted dark green, yielding a sickening smell. . . . The house was like a haunted den, long deserted by human inhabitants—deserted perhaps over scores of years" (Kim Tong-ni 1971, 60–61).

We climbed up the path to the old people's home, where the residents, ghostly pale from living indoors, tottered around to inspect us. One particularly spry old woman, her wispy hair dyed a flaming orange, danced out creakily from the dark interior, gesturing with a wave of an aqua paper fan as if she were some aged mountain fairy pointing the way: "The front gate is over there."

We went back and pounded on the gate, the right gate this time, and our knock was answered. A pleasant-faced woman led us inside to meet "Mother," the shaman shrine-keeper. The airy and immaculate interior of the house belied its decaying shell. The woodwork was polished, the floors sparkled with varnish, the door lattices were covered with fresh white paper, and a tidy garden flourished behind the house. It was the sort of Seoul house that I remembered from the early 1970s, a house that might have been built before the Korean war but upgraded with modern plumbing, running water, and appliances and maintained until the high cost of urban real estate transformed old Seoul neighborhoods into anonymous blocks of multistory construction.

The shrine-keeper, a gracious woman of almost sixty years, was, like her house, vintage but well groomed. Her round face was still handsome and she wore her hair in an immaculate old-fashioned chignon, the mark of a self-consciously traditional shaman. Her wrinkles crinkled into view when she expressed frustration at the current state of the shaman world, and she would do this often during our conversation. She was charming and intelligent and seemed happy to sit and chat with two visiting researchers on a quiet midsummer afternoon, surrounded by her apprentice "spirit daughters," all dressed as she was in loose and comfortable house clothing.

No, this was not a *kuttang*—a shrine that rents space to shamans—and this explained some of the confusion that had greeted our requests for directions. This was the shrine of a powerful local tutelary god, and it had been here for centuries. The resident spirit was a princess of the ancient Kija Chosŏn Kingdom who fled here when the kingdom fell more than two millennia ago. The shrine-keeper speculates that this powerful lady would have wreaked a series of disasters on the community until her will was heeded and her worship established. How different from the pathetic dead daughter that the old men had conjured for a shrine spirit! The ancient princess was formidable. Owing to her protection, the Manchu invaders of the early seventeenth century passed the village by as though it were invisible.[2] When People's Army soldiers commandeered the shrine during the Korean War, they all got diarrhea and promptly fled.

The shrine has been designated as a "neighborhood cultural treasure" *(tong munhwaje),* the local version of a national monument. The shrine-keeper's immediate family has maintained it and provided it with shamans for four generations. Our hostess's shaman grandmother had been favored by Queen Om, the second royal consort of Korea's last king. No, the women of this family were not hereditary shamans; they were chosen by the spirits (she uses the Korean folklorist terms for "hereditary" *[sesŭp]* versus "spiritually inspired" *[kangsin]* shamans). Her shaman mother and grandmother had both married in. The current shrine-keeper was claimed by the spirits when she was only seven years old and had been initiated as a shaman in her early teens. Her grandmother had instructed her in all of the old ways. After death, the shaman grandmother became the shrine-keeper's personal guardian spirit, as is common with shaman teachers.

The shrine-keeper evoked grandmother time to recall a purer tradition that was practiced in the twilight years of the Chosŏn dynasty, but Korea had been annexed into the Japanese Empire a full generation before her own birth. With a pained expression, she catalogued her complaints about present-day shamans who no longer tend their shrines with a proper attitude of respect, who see their sacred work as merely the means to a livelihood and shamelessly abbreviate their rituals. She deplored the current tendency for shamans to style themselves as "teacher" *(sŏnsaeng)* and "disciple" *(cheja)* or to call themselves *posal* after female Buddhist temple keepers. In dynastic times, they would say "Kijanim[3] has arrived," or call us *"sabuin."* We scribbled in our notebooks. She smiled and told us more.

Several distinguished professors of folklore had already been to her door. The names she cited were familiar to us. When she read what one of them had written, she was horrified to see how he had distorted her words.

She advises a local folklore group interested in preserving the traditions of her shrine, but their approach also displeases her. Her vexation with those who are ostensibly dedicated to recording her tradition compounds her self-image as the last of her line, a shaman who can no longer perform a proper ritual for calling the house tutelary because there are no longer any shamans who can match her chanting with the proper refrain.

Her observations are not unique. Any shaman over the age of fifty who has been practicing for more than twenty years can be counted on to express similar sentiments, and by 1994 I had been hearing these things for several years. But the setting gave this shrine-keeper's remarks an added poignancy: the old shrine and the aging shaman's deep-rooted connection to it. It had been a long time since I had sat in a tile-roofed house with lattice and paper doors. Perhaps, at middle age, I was nostalgic for the Seoul I had first known in the early 1970s—or for the Seoul I had never seen, because by the early 1970s so much of it had already been bulldozed away.

Korea is a place of displacement; few adults live where they were born, and old tile-roofed houses are found primarily in theme park "folk villages." The shrine-keeper's remembered neighborhood landscape included other shrines that had disappeared completely or were no longer adequately maintained. As new residents poured into the area after the liberation in 1945, her own shrine had been moved to what was then a desolate hillside. In the shrine-keeper's youth, wolves sometimes appeared at the summit. But her grandmother had remembered yet another move early in the last century, when the shrine was displaced by a rail line, a fitting image of modernity's assault on local religion. The former village became a part of the city, and many of the old residents—those who shared a sense of community and common history—died or moved away. The annual shrine ritual that had spiritually fortified the community through its veneration of a most Durkheimean tutelary spirit lapsed. She showed us an old snapshot of one of the last communal rituals, performed sometime in the 1960s.

Now the shrine itself was in danger. Urban planners had designated the area as parkland. All of the buildings in the neighborhood would be torn down. The shrine-keeper had already received three eviction notices but was fighting fiercely for the right to remain. She had nowhere else to go. She saw the threatened eviction as of a piece with all of the forces that had assaulted her world over the course of the twentieth century: the Christians, who called her practices "uncivilized"; the successive regimes—colonial and national—that had waged antisuperstition campaigns against the shamans; and the materialism and impatience of the current moment. She was fierce, but also pessimistic, presenting herself as the uncompromising standard-bearer of a dying tradition.

Her claims to authenticity are not what seduced my interest. I do not hold, as many students of Korean shamans do, that an older shamanic practice is necessarily "truer," "better," or "more authentic." I have always done my fieldwork in the present tense, where "tradition" is a moving target. What engaged me was the shrine-keeper's tenacity. In my mind, she was inextricably linked to her old house and through it to a nearly vanished system of local shrines, of tutelary gods in place, and of village communities that honored them. I had read about this world in the work of Korean and Japanese folklorists and was intrigued to have stumbled on its vestiges in 1990s Seoul.

When I returned to Korea four years later, Ms. Kim and I went back to the neighborhood and looked for the shrine. We had the same trouble finding it. Could it have been torn down? But this time, we had the shrine-keeper's name card, and when familiar landmarks eluded us, we placed a telephone call. She answered, seemed happy to hear from us, and sent a granddaughter down the path to meet us and guide us to the house.

The neighborhood had not become a park, but it had changed completely. The path was now lined with fancy town houses and apartments, all newly built. At the head of the path, the old tile-roofed shrine had been replaced by an imposing two-story modern structure in a traditionalist style, with shining blond wood trim and a roof of glazed blue tiles. It resembled a fancy restaurant specializing in traditional Korean cuisine. The shrine rooms had been colorfully painted in the manner of a Buddhist temple. The ambiance of a folkloric restaurant was enhanced by a display of antique earthen food storage jars, relics from her mother, displayed on an open-air storage platform as if they were still in use. The house interior was meticulously maintained as in the past but with some additions of Western furniture, including a Formica-topped kitchen table, vinyl-upholstered chairs, and a sofa. The shrine-keeper seemed renovated too. Her face was fuller and she seemed to be in a happier mood, less inclined to crinkle her face into a pained expression. Had the tension of the threatened eviction produced the pain? Even now, the issue of the proposed park was not resolved, but she and her neighbors would not have made such serious investments in construction if they had any remaining fear of eviction.

She told us again how the shrine had stood since the Imjin wars of the late sixteenth century and the Manchu invasions of the seventeenth. She elaborated on her past complaints about women who are initiated as shamans without experiencing a strong calling from the spirits. She lamented again the loss of the old ways, complaining that people become Christian to avoid the bother of honoring the ancestors, that people have become too materialistic, and that the Korean *hanbok* (traditional dress) has disappeared from everyday wear. She spoke again of how all of the old and beautiful shaman customs were disappearing and how rituals are abbreviated or not performed at all. But today, her laments for the past rang hollow as we chatted together in her totally renovated house. When she saw us out at the gate, she was again criticizing the current tendency for shamans to call each other *posal* (bodhisattva), emphatically affirming the distinction between "real" Korean *mudang,* like herself, and *posal*—the inspirational diviners who are possessed by child spirits and "scream out insults at everybody." "We used to be called *kija.* Next time I'll tell you all about it." Our exit line made a full circle back to where our conversation had begun four years before. Walking down the path, Ms. Kim and I allowed ourselves to feel let down. What we had heard before as an earnest voice from a disappearing tradition of old shrines and shrine-keepers now resembled a fixed patter, polished in conversations with others like ourselves. "When shamans sit too long with folklorists, they lose their spiritual energy," Ms. Kim observed.

I experienced these two visits as an East Asian folktale motif in reverse. In the old stories, the deluded person awakens to discover that the enchanted palace is nothing more than a crumbling ruin. In my story, disenchantment occurred when the crumbling ruin was transformed into a comfortable and somewhat ostentatious traditionalist-modern Korean house. The illusion was the product of my own nostalgia. It was the old house that had colored my view of the aging shrine-keeper. Who in 1990s Seoul would live in such a place? She must, somehow, be of another time—or at least fiercely loyal to it. But she had wanted a convenient kitchen and a vinyl-upholstered sofa after all. Was that so unusual? Perhaps the shrine-keeper's uncertainty about the public park had artificially prolonged the life of the old house. I had confused a physical structure with rootedness, material fixity, gods in place in organized space—but had it ever been so? The shrine was a portable entity. The spirits, their trappings, and the practices associated with their veneration had moved about the neighborhood; this was part of the history the shrine-keeper had shared with us on that first afternoon. Her neotraditionalist building was only the most recent in a series of three new shrines, each built in the twentieth century to replace the previous one. The "old house" that had so impressed me predated my own birth by only two years. The shrine-keeper's claim that her shrine was a neighborhood cultural treasure had nothing to do with the physical building that she had been happy to tear down and replace. She had rebuilt the shrine as a structure that signified "tradition," even as she fashioned herself as a self-conscious repository of old customs.[4]

Meanwhile, in Another Part of the City

The ultimate immateriality of the "shrine" *(tang, sindang)* in my story—the shrine as a portable name, gods, and traditions—contrasts the fixity of buildings and monuments, a contrast that may best be appreciated in a brief detour to the heart of old Seoul, which today stirs to the monumental reconstruction of the old royal capital. A half century after national independence, the imperial bulk of the building that once housed the colonial headquarters of the governor-general of Chōsen no longer blocks the view of the Kyŏngbok Palace through the (reconstructed) Kwanghwa Gate—or in a geomancer's reading, it no longer obstructs the flow of positive geomantic energy running on a north-south axis from Pugak and Samgak Mountains through the throne room of the Kyŏngbok Palace and out through the Kwanghwa Gate, "The Gate of Transformation."[5] This space of absence is as monumental in its emptiness, as much inscribed with a reading of history, as the restored palace grounds that it reveals (Nora 1989).[6] In their radical geographies, Michel de Certeau and David Harvey have drawn attention to the urban master plan as an exercise of power and domination—in de Certeau's image, a map imagined as seen from a great height, even before such heights were historically attainable (de Certeau 1984, 91–114; Harvey 1989, 70–71). The founders of the Chosŏn Kingdom understood the authority of an urban master plan when they organized their city in accord with a geomancer's reading of a power-charged landscape, and the Japanese understood it in waging architectural and geomantic warfare on that same city plan.[7] The peeling back of colonial sites to a reconstructed precolonial map replaces the material traces of the Japanese Empire with assertions of a Korean national story even as Korea's economic success, which makes such ambitious reconstructions possible, is sometimes seen as compensation for the humiliation of having been a colony.[8] These restorations of palaces, gardens, and more recently the Ch'ŏnggye Stream, fit Pierre Nora's characterization of *"lieux de mémoire"*—sites for the memorialization of "history" as distinct from a "living memory" that claims its past through repetitive and quotidian evocations. Living memory does not require monuments, and reconstructions in the name of history without literal memory are necessarily problematic and incomplete (Nora 1989). Hanyang, the royal capital of the Chosŏn Kingdom, is beyond memory's reach, but many middle-aged to elderly Seoulites caroused under the Ch'anggyŏng Gardens' cherry blossoms, trees that were tainted with colonial associations and torn up when the more sedate palace gardens were restored 1980s. Those old enough to have seen a free-flowing Ch'ŏnggye Stream—under restoration at the time of this writing—remember a fetid ditch running through a slum.

The action now shifts from a landscape of reconstructed palaces and gardens at the old city's core—already a distance from the pulse of new Seoul in the affluent high-rise districts south of the Han River—to a mountain vista and from the fixity of maps and even geomancy charts to the mutability of dreams, visions, and the agency of spirits. In my story of the old shrine, the urban planner's designation of "parkland" confronted the shaman's claims on a sacred site, a tradition, a "neighborhood cultural treasure," and we will see equivalent forces repeatedly at odds throughout my discussion. Disputes over shrines, real estate, and urban parkland are informed by contradictory, simultaneous, and interpenetrating perceptions of "landscape"—landscape in its emergent anthropological sense as "the meaning imputed by local people to their cultural and physical surroundings (i.e., how a particular landscape 'looks' to its inhabitants)" (Hirsch 1995, 1; see also Basso 1996; Munn 2003).

Caroline Humphrey (1995) has analyzed Mongol history as a shifting ascendancy of maps and shamanic landscapes, the one fixed, legible, and consequently finite, the other infused with the more fluid stuff of the imagination—including the agency of spirits. Her characterization is useful in thinking about confrontations between Korean shamans and urban planners, but where Humphrey posits a diachronic relationship of alternately dominant visions, the shrine-keeper of my tale and the authorities that had issued her eviction notice were engaged in a synchronic and unequal contestation. In Barbara Bender's words, "The landscape is never inert, people engage with it, re-work it, appropriate and contest it. It is part of the way in which identities are created and disputed, whether as individual, group, or nation-state" (Bender 1993a, 3; see also Bender 2002a,b; Bender and Winer 2001; Kuchler 1993; Kuper 2003; Morphy 1993). Borrowing on the theme of the urban walker, from Baudelaire through Walter Benjamin, Michel de Certeau describes the totalizing master plan of a city as confounded and reimagined by "walkers" on the ground whose everyday practices and "superstitions" are "foreign to the 'geometrical' or 'geographical' space of visual, panoptic, or theoretical construction" (1984, 93, 106). Might such reimaginings not also be the work of shamans, spirits, and shrines?

Mountain Shrines

Nestled on the slopes of Seoul's mountainous periphery are sacred sites: spiritually potent rocks, springs, *sŏn'ang* trees that harbor unquiet ghosts, and—for the last twenty years or so—a proliferation of *kuttang*, a few with significant histories, but most of them having sprouted up like mushrooms after the spring rain. Built on quiet and relatively isolated hillsides, several of these *kuttang* are already surrounded by ballooning residential neighborhoods. *Kuttang* rent space for shamans' *kut*—ceremonies that entertain the spirits with percussive music and song, dancing, mime, and comic play as the costumed shamans embody the gods and ancestors of their clients. Here, room by room, different regional styles are performed simultaneously, their distinctive rhythms and stories spilling into the common courtyard in a montage of action and sound, a postmodern ritual happening. The grounds of some *kuttang* include additional sites for prayers to the Mountain God (Sansin) and the Seven Stars (Ch'ilsŏng) of efficacious mountains *(myŏngsan)*. Many efficacious sites, such as the Pohyŏn Sansingak on Samgak Mountain, were known to worshippers long before the appearance of *kuttang* bearing their name (cf. Yi Chae-gon 1996, 162–164). Others are newly constructed with wells of electrically pumped water to simulate mountain springs and cement platforms for offerings.

Koreans sometimes remark that *kut* in villages and even urban neighborhoods were a relatively frequent spectacle until sometime in the 1980s. Some even assume that with the disappearance of *kut* from public view, shamanic practices have died out. This is belied by the contemporary flowering of mountain *kuttang*. The spread of *kuttang* has been linked to antinoise ordinances and "antisuperstition" campaigns in the 1970s, all blatantly intended to suppress the shamans' activities (Huang 1988, 18; Kim Seong-nae, personal communication 2003; Sun 1991, 163). Cramped apartments and life lived on industrial time further enhanced the appeal of the relatively isolated *kuttang*, where there was space for offerings, dancing, and feasts—and no neighbors to complain about the noise. With rising standards of consumption, shaman households acquired vans to transport clients, offering food and ritual paraphernalia to mountain *kuttang*. From the early 1990s, "convenient parking facilities" have been advertised on *kuttang* business cards. All of these developments have had a profound impact on the pacing and tenor of *kut,* making them abbreviated and private events compared to the village or neighborhood spectacles of a few decades past.

Kuttang, like inns, restaurants, and wedding halls, are service enterprises. In addition to renting rooms with altar space and an appropriate configuration of god pictures, *kuttang* staff provide vessels for the offering food, cook the offering rice, steam the rice cakes, cater meals on request, and vend cigarettes, drinks, and tonics at inflated prices. Like other service enterprises, *kuttang* compete to offer upgraded facilities such as indoor plumbing, air conditioning, and expansive parking space. One *kuttang* I visited was equipped with an electric dumbwaiter for hauling offering food to shrine rooms on the upper floors of the multistoried building. Some *kuttang* provide costumes and equipment for young shamans who have not yet acquired their own. *Kuttang* personnel know how to arrange offerings, anchor a full pig's carcass on a trident, and prepare fodder choppers for a shaman to balance on. While some proprietors of *kuttang* have the air of innkeepers or restaurateurs, others are shamans themselves or have strong links to the spirits whom they must necessarily honor, on a daily basis, as shamans do. Shrines are gathering places for those who feel a close association with the spirits but are not quite shamans themselves, musicians *(chaebi)* who often come from shaman families, and maids-of-all-work *(kongyangju)* who are destined to be shamans but lack either money or sufficient inspiration for a successful initiation ritual.

Kuttang are both a new and an old phenomenon, a bringing together of two discrete phenomena: shrines dedicated to significant tutelary spirits (*tang* but not yet *kuttang*) and pilgrimages to sacred and potent mountains *(myŏngsan)*. Seoul's most venerable *kuttang*—the Kuksadang, originally located on South Mountain (Namsan)—was one of several tutelary shrines that were a part of the Chosŏn capitol's sacred landscape. *Pugundang* were shrines to spirits associated with different government offices and magistracies and were maintained by the clerks (Walraven 1995, 112; Yi Hŭi-sŭng 1972, 1,283). Potent local deities, like the maiden of the old shrine, were also granted official recognition and housed in *pugundang* (Pak 2001; Yi Chae-gon 1996; Yi Nung-hwa [1926] 1976). With the passing of the kingdom, local shrines lost their association with any national administration, and their periodic rituals came to be regarded as village festivals, in most instances maintained and perpetuated by shamans, as in the tale of the old shrine. Several *pugundang* along the Han River, now embraced by the sprawling boundaries of metropolitan Seoul, remain sites of ritual activities and folkloric inquiry (Pak 2001).

The historical connection between official shrines, shamans, and the holding of private *kut* is unclear. The present-day keepers of the Kuksadang maintain that in the past, only high officials participated in the rituals that shamans performed there. An early-twentieth-century observer records, however, that at the time of his observation the most impressive and lavish *kut* were performed in public shrines where, in addition to the cost of offering food, "a certain sum must be paid as rental" to "the owner," probably the shrine-keeper (Korean mudang and pansu 1903). Walraven suggests that this was a recent development around 1900 and that the vehemence of government campaigns to remove shamans from public shrines in the waning years of the Chosŏn Kingdom was inspired in part by the impropriety of shamans holding private *kut* in these places (1995, 129; 1998, 56–59). Be this as it may, the anonymous early observer records a significant list of shrines that accommodated private *kut*. Of these, the Kuksadang, the Sasindang, and the Halmidang were well known to researchers in the 1970s when I did my first fieldwork, and the Kuksadang and (relocated) Sasindang still function today. The Kuksadang seems to have assumed the appearance of a contemporary *kuttang* from at least 1925, when it was removed from South Mountain to its present site on Inwang Mountain and its management passed into private hands.

In the 1970s, most of the *kut* that I attended were performed in private homes or in the shamans' own personal shrines, as they had been in late dynastic times (Korean mudang and pansu 1903). Mountains were barely accessible pilgrimage sites that shamans and ordinary women would visit occasionally, sometimes with great difficulty. Midway between earth and sky, mountains are concentrated domains of spirits and consequently potent sites for prayers, places that all shamans visit periodically to "recharge" their inspirational batteries (one of my informants used this metaphor himself) and where the prayers of even ordinary people are particularly efficacious.[9]

Although in the 1970s many of the mountains near the capital housed military observation posts, their association was as wild, uninhabited, magical space where the spirits' force was strongest and the shamans' visions the most vivid. I accompanied Yongsu's mother on a daylong pilgrimage to the summit of Kambak (or Kam'ak) Mountain in 1977, and nearly ten years later—when the summit had been placed off limits—on a dawn trip by taxi to the foot of a path leading to the shrine tree on a lower slope of that same mountain. In the early 1990s, she began to take her clients in the family car (most recently an SUV) to her favorite among the several *kuttang* that have been built on the lower slopes of Kambak Mountain. The *kuttang* kitchen steams the rice, water is drawn from a cement-encased well, offerings are arranged on one of several outdoor altars, mats are provided for prostrations, and one can go and return home in under three hours. Still, these mountain prayers are serious acts requiring three days of vegetarian fast and a state of ritual purity for both shaman and client beyond what would be required for a *kut*. Living in a quasi-rural setting, Yongsu's mother performs *kut* for her own clients in her private shrine, but I have accompanied her to many other shamans' *kut* held in newly constructed *kuttang* in the mountains surrounding Seoul and Suwŏn.

The contemporary *kuttang* thus blends the use of shrines for *kut* (some located on mountains, some on particularly potent mountains) with the destinations of mountain pilgrimages (which now often utilize space maintained by commercial *kuttang*). On mountain sites, *kuttang* are also nodes of connection between shamanic notions of mountains as places infused with spiritual and visionary power and the geomancer's scheme of a systematically magical landscape charged with invisible veins. Particularly auspicious is the level space below a potent mountain configuration—the *myŏngdang*—where the positive energy of the site converges (Kim Yong-ok 2003). Like any business site (or home or grave), a *kuttang* benefits by being situated on a *myŏngdang* not only for the success of the enterprise but to the benefit of all who avail themselves of the shrine's services (the rumored efficacy or inauspiciousness of a site will, of course, affect business).[10] *Kuttang* that intersect with powerful mountain geomancy can claim this as a selling point; one *kuttang* that I visited southeast of the city advertises itself as basking in the energy flow *(maek)* of three famous mountains on the surrounding skyline. But while geomancy's energy flows can be predicted and mapped, the spirits assert their presence in the landscape in surprising ways.

Shrines, Gods, and Power

One morning in 1994, I found the young attendant of a small *kuttang* glumly surveying the wreckage of a metal frame and nylon canopy, a tentlike structure intended to shield supplicants at a sacred well. The neighborhood authorities had torn it down because the *kuttang* sits on officially designated parkland where building is prohibited. The *kuttang* itself was safe, he told us, because it had been registered as a shrine for more than twenty years, but the payment of "fees" had not been sufficient to spare the new structure. This small encounter—combined with the story of the old shrine and other stories to come—is indicative of the spatial conflicts and ambiguities that conspire around shrines, conflicts that may be as old as a century of Korean modernity (Walraven 1995, 1998). Even without the blunt instrument of "antisuperstition" campaigns that sanctioned the destruction of village shrines in the early 1970s (Ch'oe 1974), master plans for "urban renewal," "parkland," and "security areas" have dislocated shrines and placed many sacred sites off limits to worshippers (Pak 2001; Ryu 2000). Je-Hun Ryu (this volume) offers evidence of quiet but persistent resistance. Pak Hŭng-ju's survey of neighborhood shrines *(pugundang)* catalogues their physical displacement by the construction of new roads, highways, and apartment blocks, the subsequent dispersal and dying off of old communities of adherents, and the tenuous perpetuation of old rites—the keeper of the old tile-roofed shrine's complaint in multiple (Pak 2001).

The Granny Shrine (Halmidang) grew up around a shrine tree at the northwest pass on the highway out the city of Seoul. In one version of the legend, tigers used to prey on travelers navigating the pass until a wise old woman related that the spirit in a tree beside the path was demanding offerings. The tree became a *sŏn'ang* shrine, and the wise old woman was eventually apotheosized as the "granny" of the shrine. The shrine outlasted the tree, which was cut down in colonial times, possibly as an "antisuperstition" measure; according to oral tradition, it fell on the man who wielded the ax, crushing him to death. The teller gave the tale a nationalistic spin, of a piece with geomancy stories abroad in Korea at the turn of the millenium: The Japanese cut the shrine trees down to make telephone poles and the Japanese cut the auspicious veins of Korea's geomancy with iron bars and axes; Korea's subsequent troubles—national division, war, the Asian financial crisis—stem from these maleficent interventions on its landscape (Kim Seong-nae 2000). The Granny Shrine outlived the shrine tree but did not survive more recent complaints of noise pollution from Christians in the neighborhood; by 1990, it had disappeared.

I heard the legend of the Granny Shrine from the keeper of another old shrine, the Fortification Shrine,[11] originally in the same neighborhood, which her family had managed for three generations. The story of this second shrine is a tale of more recent but no less fantastic spiritual agency. Four years before, the current shrine-keeper's brother had sold the shrine to settle a gambling debt. She did not tell us about the debt; Ms. Kim and I had heard about it in the sideline gossip at another *kut* that same summer.

What she did relate, lowering her voice for Ms. Kim on the assumption that her story was not fit for a foreigner's ear, was that within a hundred days of closing the deal, all three parties to this transaction—the man who had made the introduction, the man who fronted the money, and the new manager—were killed in a traffic accident, widely interpreted as a punishment from the spirits.[12] This, too, the foreigner had already heard from the musicians at another *kuttang*, but the shrine-keeper brought the story up to date. The deaths cast a pall over the enterprise, and the shrine was torn down and replaced by an ordinary building with no connection to the shaman world. The shrine-keeper claims that she is the only member of her family to take the spirits seriously, the only one willing to reestablish the Fortification Shrine in a new (and less expensive) location, and she has persisted, despite her husband's objections. She received a dream from the spirits telling her to look for a new site in a hilly northern neighborhood of the city. To her surprise, she received a telephone call from a potential backer within days of this auspicious dream and had soon set herself up in business in an old house near a Buddhist temple. Here she reinstalled the god images and paraphernalia from the old shrine and maintained the name. When we met her, she had been in this new location for about four years, but business was very slow that summer and her only regular clients seemed to be the old shamans who had patronized the shrine in her father's day. In her present location, she could not meet the rising standards of clients who expected modern plumbing and parking space.

In the summer of 1998, Ms. Kim and I went back to the Buddhist temple wondering if the shrine proprietress had managed to sustain her business in the precarious economic climate following the Asian financial crisis. The Fortification Shrine was gone, replaced by a cinder block house, an ordinary house with a pile of shoes by the door and a child's plastic toys abandoned in the front yard. No one was at home. The proprietor of the snack shop down the road remembered *kut* in the neighborhood; he told us they had stopped "long ago." We had watched *kut* at the Fortification Shrine only four years past, but he may as well have been recalling events from his childhood. The monk taking tickets at the temple had no memory of the *kuttang*, but young monks are not long in one place. Another monk, overhearing our conversation, told us that the proprietress had moved away. We would hear later that the temple had not been happy with a *kuttang* at its front door. As a final gesture, we called the telephone number on the shrine-keeper's old business card and, to our surprise, she picked up! She had relocated to the far north of the city, once again taking the old shrine fittings and shrine name with her. The Fortification Shrine was now located beyond the (then) northernmost extension of the orange subway line in a place of fields and trees. It was a spacious shrine in a large tile-roofed house—four shrine rooms and a broad parking lot. The shrine-keeper's face had softened, the anxious lines were gone, and things were going well for her. The old *kuttang* had been "someone else's house," she said, but this one she owns. Her clients had loaned her the money to buy and repair it. She had been here for two years, having moved in on the ninth of lunar September, "the day the swallows fly south of the Han River."

Once again, the location of her *kuttang* had been foretold in a dream. In the dream, her junior uncle and aunt, described as "the parents who had raised her," and one of the spirit "grandfathers" in her shrine led her to a lotus pond with a waterfall cascading down. She thought that she could not pass through the curtain of water, but somehow she managed. She was standing with her uncle in some house—this very house—and her uncle put an official certificate into the storeroom above the kitchen. She recognized the house from her dream when she came here for the first time, and she learned from the neighbors that there had been a lotus pond nearby in the past. In her dream, she had felt a need to relieve herself. A male and a female god led her to a place where there was water pouring down. She was reluctant to relieve herself there, thinking of the people downstream, but the male god told her, "I'm the president and I've set my seal to it." Complaints from neighbors—the bane of shamanic activities—were an expressed concern of hers. Recalling the fate of the Halmidang and her own difficulties with the Buddhist temple, perhaps the spirits in her dream were reassuring her about the viability of her *kuttang* in this new neighborhood.

The gods' agency in selecting the place where they will reside is a recurrent theme in the dreaming of shamans and shrine-keepers. Pak Hŭng-ju's survey of *pugundang* along the Han River provides several historical and recent examples of shrines that had been relocated with the aid of dreams (2001, 108–109, 167, 182). Yongsu's mother described how, earlier in her career, her gods had insisted that she install them in the spare room that she was renting out to strangers. "We'll give you the rent money," they told her, and subsequently they made good on their promise by bringing her clients and giving her a successful practice (Kendall 1985, 56). Years later, in a splendid new house, her gods insisted that she move their shrine from a side room to a more central location; the side room was too much like a servants' quarters, located on the side of the outer courtyard in a traditional Korean house. Yongsu's mother had been afflicted with bad business and a sore leg until she realized her mistake.[13] The Fairy Maid, a young shaman whom I met in 1994, claimed that her "grandfathers" had helped her to secure an ideal site for her home and personal shrine under very favorable rental terms.

The Celestial Shrine is a popular and well-appointed *kuttang;* in the early 1990s it claimed patronage by Korea's premier National Treasure shaman, and in the hot summer of 1994, this was the only shrine where we found air-conditioning. The proprietor happened upon this site after seeing it in a vision during a mountain pilgrimage. We heard his story in 1998 when he had just returned from another pilgrimage to various mountains to revitalize his spiritual energies, a task well worth the effort since he had returned to find preparations for a *kut* underway, welcome business in that difficult spring when the shrines, like many businesses, were feeling a sharp falling off of business in the wake of the so-called IMF (International Monetary Fund) crisis. Perhaps the unforeseen business put him in an expansive mood (he had refused to speak with us at all in 1994), and he spoke of many things, including the story of how he had found the location for his shrine. He had gone to T'aebaek Mountain, far to the south, to pray at a particularly potent site and there had a waking vision *(hwansang)* of a plateau surrounded by mountains, a configuration resembling an old woman sitting on her haunches (he struck a pose to illustrate). It was exactly this place, a sacred site where women came to pray, and here he built the Celestial Shrine, naming it after the site in the T'aebaek Mountains where he had experienced his vision.

But however divinely decreed, in 1998 the Celestial Shrine's days seemed to be numbered. A tunnel was being cut through the mountains, and the surrounding area had been declared parkland. The Celestial Shrine was scheduled for demolition within the next two years, along with several other shrines in the immediate neighborhood. We asked him where he would go next. He did not know; he planned to establish a small retreat somewhere in the mountains. It seemed as though he would wait for another vision. Four years later, it was nearly a replay of the story of the old tile-roofed shrine. We found the shaman shrine-keeper—and his neighbors—still in business. He was entertaining his cronies on a busy Sunday. When I reminded him that the last time I had seen him it had seemed that the shrine would be torn down, he smiled as at a distant memory: "Oh yes, there had been that talk."

When a *Kuttang* Becomes a Heritage Site

Articulate shamans and representatives of shamans' advocacy organizations describe the harassment of *kuttang* and the closing off of sacred mountain sites through a new language of religious discrimination, claiming that Buddhist temples and Christian churches would not be subject to arbitrary demolition. Some shrine-keepers have begun to deploy a language of heritage sites as a protective strategy. But as I learned from the demolition and reconstruction of the tile-roofed shrine, a language that equates shrines with buildings misses their primary claim as supermaterial gathering places for spirits. With the end of the Chosŏn Kingdom, the oldest shrines in this discussion were unhinged from the geomantic scheme of the old royal capital and have wandered from their original sites. The Fortification Shrine, relocated twice in the 1990s after having nearly been lost in an inauspicious real estate deal, had already been relocated once in the 1960s when the north-bound road was widened in an early spurt of urban development. The "old shrine" of my opening story had been displaced twice—first by a railroad, then by urban sprawl. Korea's most famous *kuttang,* the Kuksadang, carries a well-known story of colonial displacement. Originally located on South Mountain, the southernmost extension of a geomantic configuration that begins with Pugak Mountain in the north, creating an axis through Kyŏngbok Palace, was displaced in 1925 by a Shinto shrine in a further colonization of the Korean landscape (Chŏng Chong-su 1999; Kim Yong-ok 2003). The Kuksadang was relocated on Inwang Mountain to the northwest, a mountain also significant in the city's geomancy.[14]

As a site of living shamanic practice and a material monument to dynastic memories and colonial displacements, the Kuksadang bridges the divide between a functioning *kuttang* and a material monument, an exception that by its exceptionality highlights the contradiction between shamanic versus mapped landscapes, between functioning shrines and monumental *lieu de mémoire.* A national heritage site since 1979,[15] the Kuksadang's example may have inspired the old shrine-keeper to post a signboard describing her shrine as a "neighborhood national treasure" and the keeper of the Fortification Shrine to describe her once more newly relocated shrine as an "important folk property." Within the Korean shaman world, claims of folklorists and cultural historians for a "deep cultural root" of shaman practices have won for shamans a measure of public recognition and have led to the appointment of some shamans as human cultural treasures *(ingan munhwaje).* These developments are generally seen as both preserving endangered cultural practices and enhancing the self-esteem of the practitioners. It follows that official designations of monumentality might confer respect, bestowing a mantle of protection upon those precarious shrines that claim genealogies dating back to dynastic times. In the case of the Kuksadang, however, heritage status has been a mixed blessing.

It would be difficult to find a shaman or a folklorist who has not heard of the Kuksadang. An important national shrine in dynastic times, the Kuksadang is widely regarded as Korea's premier *kuttang.* Among traditionalist circles of Seoul shamans, doing a *kut* at the Kuksadang confers the status of a fully realized professional shaman.[16] Inwang Mountain, where the shrine is located, is important to the geomancy of Seoul as the "white tiger" in the configuration of mountains protecting the capital from the north (Kim Kibin 1993), and the *myŏngdang* on which the shrine is located is said to be extremely powerful. The Sŏnbawi, a large twisted rock near the shrine, is considered a potent locus of prayers.[17] Lay people come here to pray for conception and good fortune, and shaman initiates pray for the visions and inspired speech that will make their initiation a success. Inwang Mountain is a power-charged landscape, and with the end of security restrictions and the opening of the mountain to hikers during the Kim Yŏng Sam administration, people have come to pray all over the mountain, often leaving piles of trash and rotting offering food in their wake. This activity provokes the ire of the many Buddhist establishments that have also grown up in recent years on Inwang Mountain. In this now crowded space, the interests of the monks are pitted against the activities of shamans who, to monks' eyes, pollute the mountain and interrupt meditative practices with the clanging of drums, gongs, and cymbals. Noise complaints are a major issue for the Kuksadang, where *kut* must end by the late afternoon lest the monks from one of the surrounding hermitages complain, once again, to the authorities. The Buddhist establishment has assumed the management of the Sŏnbawi, improving the site but prohibiting the initiation *kut* that used to take place on the narrow ledge in front of the spiritually potent boulder.

The aura of the Kuksadang's national heritage status does not seem to have conferred any particular privilege on this shrine in the micropolitics of Inwang Mountain. In 1994, a signboard just outside the Kuksadang proclaimed the authorities' limited tolerance of *kut:* "In accord with official policy we beseech you to abridge your activities, refrain from drinking and rowdy behavior, and conduct your work in a dignified manner, scrupulously adhering to the designated time." In 2003, a sign proclaimed that "Prayers and shamanic activities *[musok haengwi]* and the like are prohibited within the Inwang Mountain Municipal Nature Park." This official decree has had no visible effect on activities either inside the Kuksadang or all over the mountainside.

Policies intended to protect the old wooden structure of the shrine are at cross-purposes with its role as an active *kuttang*. The lighting of candles, an important element of *kut* offerings, is technically prohibited. In the summer heat, the shrine-keeper carefully monitors the electric fans set up in the shrine for fear that overheated wiring will produce a fire. It is not possible to install air-conditioning in the shrine because efficient insulation would require modifying the doorframes, violating the original architecture. The shrine-keepers are also caught in a contradiction that besets the owners of heritage property in other places (Herzfeld 1991). As private owners, they are expected to maintain the building, covering necessary repairs out of pocket, but all repairs must be carried out using approved traditional methods and materials. State-of-the-art authenticity is often very costly. As a resolution to this contradiction, there is talk of the Bureau of Cultural Properties Preservation taking over the shrine and maintaining the old building as a historic site. What, then, would it represent? As one more *lieu de mémoire* of old Seoul and of the ruptures imposed by its colonizers, would it become, as in Barbara Bender's critique of Britain's Stonehenge, "a socially empty view of the past" (Bender 2002a, 169)? However it might speak to "history," the Kuksadang would cease to be an active *kuttang*. Devoid of gods, visions, and shamanic performances, it could only be a monumental shell. The gods, presumably, would go elsewhere.

Conclusion

One can read the peripatetic histories of shrines as I initially read them: as reactive responses to the hegemony of maps, state projects, and the encroachment of disenchanted real estate. But disenchantment is seldom absolute, and perhaps the spirits are innately resilient. If the veins of a geomancer's map can be strangled with iron bars, destroying the potency of a site, the spirits seem to have a regenerative capacity to inhabit and reinscribe new landscapes with "the magical power of proper names" (de Certeau 1984, 104). Buildings may crumble, but spirits are not one with the material substance of shrines. Shrine-keepers and shamans proudly tear down and enlarge their own shrines, assuming that the spirits, like mortal Koreans, take pleasure in more modern and spacious accommodations. "Old shrines" persist in the names they bear, the gods they house, and the recognition of practicing shamans, and most *kuttang* are not very old at all. With other commercial enterprises, they are free to wander the landscape. *Kuttang* make their claims on the landscape through the unmappable and unpredictable agency of spirits as revealed in dreams, visions, shamanic performances, and extraordinary happenstances. In Nora's terms, this is the stuff of "memory," not "monument." In de Certeau's scheme, these are the "superstitions" that muddle and confound a totalizing city plan. Popular imagination keeps alive small but fantastic acts of magical resistance: The Granny Shrine's sacred *sŏn'ang* tree fell on and crushed whatever agent of modernity had cut it down; the businessmen who brokered the Fortification Shrine died in a traffic accident. As unruly memory sites, the *kuttang* function as de Certeau's "anti-museums," deploying the stuff of legend against a totalizing vision of urban space, offering the possibility of escape "into another landscape" (de Certeau 1984, 106–108), albeit not one of picturesquely crumbling tile-roofed houses but like much of the urban periphery, hastily constructed of cheap and expedient materials because it will be torn down and reconfigured soon enough.

Notes

This essay draws primarily upon short research trips to Seoul in 1994, 1998, 2002, and 2003, which were supported by the Jane Belo-Tanenbaum Fund of the American Museum of Natural History. I am grateful to Ms. Kim Sung Ja for research assistance and insight and to the many shamans and shrine-keepers who were willing to talk to me. I am also grateful to Dr. Kim Seong-nae and Dr. Chŏ Chonsu for providing me with important bibliography, including their own fresh approaches to the geomancy of Seoul. I alone am responsible for the shortcomings of this effort.

1. An early version of this story was originally published in the journal *Shaman* (Kendall 2002).

2. She gives a complex reason for the tutelary god's effectively repelling the Manchus. The tutelary god was a refugee princess who fled to this place with the fall of the ancient (legendary) kingdom of Kija Chosŏn, purported to have fallen two centuries before the common era. The shaman repeats popular historical understandings of a "Korean" kingdom (Koguryŏ) that had once dominated Manchuria. A Kija Chosŏn princess could thus subdue (or at least hoodwink) a barbarian army, the putative former subjects of her house.

3. The derivation of this term is unclear. She attributes it to her tutelary spirit, a princess of the ancient royal family of "Ki." I have heard it used among older shamans, but not by their clients.

4. This shrine-keeper may be the niece and granddaughter of two shamans interviewed by Youngsook Kim Harvey thirty years ago in this same neighborhood (Harvey 1979, 17–36). If so, then the shrine-keeper's posture was a family tradition. Harvey describes these shamans as exuding "an aura of legitimacy that is absent for the most part from the other shamans I have met" (19), and they were the frequent informants of folklorists. I am grateful to Roger Janelli for reminding me of Harvey's account.

5. For more about the destruction of the governor-general's headquarters, see Kim Hong-nam (1998, 183). For accounts of its history—which in its last incarnation was the National Museum of Korea—and debates surrounding its demolition, see Pak (1997) and Kim Hong-nam (1998). For an account of how these events adumbrated in the popular historical imagination, see Kim Seong-nae (2000).

6. The colonial reconstruction of Seoul and recent attempts to reclaim a historically Korean landscape are described by Henry and Jin (both in this volume).

7. See Chŏng Chong-su (1999) and Kim Yong-ok (2003) for accounts of the geomantic principles involved in the citing of Hanyang as the Chosŏn Kingdom's royal capital. That both of these works were written for a general audience suggests the appeal of geomancy talk in the contemporary imagination. Chŏng is particularly good in recounting legends of the power struggle between the Buddhist sage Muhak and the Confucian scholar Chŏng Tojŏn, enacted through their differing opinions on how the built capital should be positioned in relation to its geomancy.

8. See Cho Hae-joang (1998) and Chungmoo Choi (1993b) for discussions of the problematic nature of South Korea's postcolonial national identity, which Choi describes as a failed project and Cho, writing at the end of a decade of economic success, sees poised for takeoff. Kim Seong-nae (Sŏng-rye) (2000) links the geomancy stories associated with the demolition of the old colonial administrative center to the lingering problem of a colonized modernity.

9. I have written about mountain prayers and pilgrimages in several different places (Kendall 1977, 1985, 1988, 1998).

10. The notion that an enterprise's site has a contagious effect on all who use it is very widespread. In urban legends, wedding halls fail when word gets out that their weddings end in divorce, and hotel coffee shops are reputed to be either particularly auspicious or particularly inauspicious for the formal first meeting *(massŏn)* of a prospective bride and groom.

11. I have tried to at least thinly disguise the identities of the shrine-keepers who were my interview subjects, the one exception being the Kuksadang, which is central to my story in ways that are impossible to disguise. In this instance, I have been even more cautious than usual in using this interview material.

12. Pak Hŭng-ju (2001) records a similar story about another shrine owner and broker who met bad ends owing to their profit-motivated sale of a shrine.

13. They had also complained that she did not wear traditional padded socks when she worshipped them, a further sign of disrespect to very noble spirits such as themselves.

14. The rights to the shrine fittings and the name were purchased by a private entrepreneur whose family manages the shrine today.

15. The paintings inside the Kuksadang, the oldest known shaman paintings in Korea, received National Treasure status in 1974.

16. The keeper of the Kuksadang and other informants have described—always in the past tense—a circuit of shrines associated with potent mountains that shaman initiates were required to visit to secure the gods' favor. The descriptions of the circuit are inconsistent, but all culminate in the Kuksadang.

17. The lure of the potent Sŏnbawi figures in Yongsu's mother's story of her shaman destiny (Kendall 1988, 73) and in some other stories of destined shamans' experiences.

7

Kyeryong Mountain as a Contested Place

JE-HUN RYU

KYERYONGSAN (KYERYONG MOUNTAIN) rises alone out of the otherwise flat countryside that surrounds it, giving it a somewhat mystical appearance. Numerous peaks form an elongated ridge resembling a dragon with a rooster's comb—not surprisingly, the name "Kyeryongsan" means Rooster Dragon Mountain. The striking shape of the mountain has captured the imagination of people for years. Along the peaks and in the deep valleys that cut across the mountain, many places lure both shamans and their clients. Today it is known in the area as the most important mountain for shamanists. They visit it to pray and perform ceremonies. These shamanists usually stay for long periods—several days or longer—strengthening their supernatural abilities and, at times, establishing their own halls.

Yet Kyeryong Mountain is not the sole domain of shamanists. Different religious or ideological groups have endowed the space with amalgams of meanings, uses, and values. Shamanists, Buddhists, and Confucian followers, along with groups engaged in various religious philosophies, have used the mountain in their practice to create and maintain territories and identities. At the same time, the state has proposed narrow uses for the mountain that at times have been directly hostile to one or more of these other groups, who also have laid claim to and redefined the mountain. Interestingly, it appears that shamanists—possibly the most marginal of all these religious or ideological groups—have managed best to resist domination from the state, first sidestepping the government's designation of the mountain as a national park and then its designation as a military headquarters. Through all this, the shamanists have managed to maintain their control of the mountain through the practice of spatial tactics.

State control and its spatial implications for other users of the mountain are nothing new—rather, the state and its practice of control have intervened in the territorializations of the mountain throughout time. The religious or ideological groups have presented various forms of resistance to each other and the state. The differences in the power relationships between the state and these various groups have also given rise to conflicts over the uses of space. Within such contradictory uses and interpretations of the mountain space, particular locations on Kyeryong Mountain have become sites of contestation where power relations are interwoven. Consequently, these sites align with what Lefebvre has described as the alternative construction of place, or "counter spaces" (1991, 382).

While the central strategy of authority is to force people to acquiesce, people often resist. Pile, in his elucidation of geographies of resistance, proposes that people have the capacity to change the official definition of place through the deployment of tactics that allow them to avoid, taunt, attack, undermine, hinder, or mock the spatial exercise of power of the dominant group (Pile 1997, 14; see also de Certeau 1984). Resistance reflects the desire to find a place in a power geography where space is denied, circumscribed, and/or administered. For example, followers of the *Chŏnggamnok* and other related belief systems can be seen as marginal groups who were anxious to find a space for their religious practice during a time of political turmoil—namely the Japanese occupation of Korea (1910–1945).

The spatial intersection of the powerful who control space and those who deploy tactics of resistance has at least two "surfaces": one facing toward the mapping power and the other facing in another direction, toward intangible, invisible, unconscious desires, pleasures, enjoyment, fears, angers, and hopes. On Kyeryong Mountain, the powerful have been the state, with its imposition of a national park and military headquarters onto the space of the mountain, and the weak have been the shamanists and the followers of other popular religions. The Buddhists, interestingly, have perched on a fine wire between power and resistance. In a geography of resistance, secondary places are always involved: spaces that are dimly lit, opaque, deliberately hidden, or saturated with memories (Pile 1997, 16). The spaces of resistance seized by the shamanists have been simultaneously multiple, dynamic, and weak—always only partially controlled by the practices of domination.

Kyeryong Mountain provides an ideal case study for thinking through geographies of resistance and entanglement. Geographies of resistance seek to occupy, deploy, and create alternative spatialities from those defined through oppression. Through history, marginal and dissident groups on Kyeryong Mountain have transgressed the authoritarian imposition of official boundaries to survive that oppression (and its attendant systems of surveillance), first from neo-Confucian rule and later from the military government. The aim of this brief study is to examine how shamanists, Buddhists, and other religious groups have sought to occupy new spaces, to create new geographies, and to make their own places in opposition to the official projections of what Kyeryong Mountain should be. As part of this project, an attempt is made to unveil how these marginal groups occupied the Sindoan Valley—and other smaller places—from the early 1900s and how they reacted to the construction of a military headquarters on the mountain in the 1980s.

Local History of Confucianism and Buddhism

Ch'ŏnhwang (Ch'ŏnhwangbong, 845 m) is the highest peak on Kyeryong Mountain, and from it ridges radiate in five directions, all with lesser peaks defining their ridge lines: Ssalgye (828 m), Kwanŭm (816 m), Sambul (775 m), and Sujŏng (662 m) in the north; Sinsŏn (642 m) and Changgun (500 m) in the northeast; Ch'ŏnhwang (605 m) and Hwangjŏk (664 m) in the east; Hyangjŏk (574 m) and Kuksa (431 m) in the south; Munp'il (816 m) and Yŏnch'ŏn (739 m) in the west. Waters flow from the summit, forming three large streams—Yŏnsan, Tugye, and Nosŏng—eventually draining onto the plains below the mountain. The mountain itself is divided into five valleys. In this study, these five valleys are named tentatively after representative cultural landscapes that define each of them: Tonghaksa, Sangsilli, Sindoan, Kapsa, and Sinwŏnsa.

Kyeryong Mountain has been considered sacred by the state from the earliest periods of Korean history. In 420 CE, the Paekche Kingdom (48 BCE–660 CE) built the first Buddhist temple, Kapsa, on the mountain. The *Samguk sagi* (History of the Three Kingdoms) (Kim Pu-sik 1998) mentions that five sacred peaks were identified during the Silla Kingdom (57 BCE–935 CE) to protect the nation. Among these was Kyeryong Mountain, called at the time Sŏak, "the sacred mountain in the west." In 1394, the royal court established an altar at the temple of Sinwŏnsa to pay homage to the deity of the mountain. From that point on, Kyeryong Mountain remained as one of the three sacred mountains where the royal court paid homage to the mountain deities.

In 1393, the founder of the Chosŏn dynasty, King T'aejo (reigned 1392–1398), visited Kyeryong Mountain with the idea that he would relocate the royal capital from Kyegyŏng (currently Kyesŏng) to the Sindoan Valley. Although he changed his mind and located the royal capital in Hanyang (currently Seoul) instead, Kyeryong Mountain came to be known as the most sacred mountain in Korea. In the mid–Chosŏn dynasty, a book of geomantic prophecy called *Chŏnggamnok* became popular among those who were dissatisfied with the current sociopolitical situation. The book said that the Chosŏn dynasty was coming to an end and that a new dynasty would succeed it. Its capital would be in the Sindoan Valley on Kyeryong Mountain (Yi Min-su 1993, 13).

With the Japanese annexation of Korea in 1910, adherents of various popular religions moved to the valley, as they believed that it would become the center of the nation—or perhaps even the world—in the not-so-distant future. After the Korean War (1950–1953), shamans and their clients came to Kyeryong Mountain in large numbers both to pray and to perform rituals. But in 1975, the Korean government created a national park covering much of the mountain, and from that time on the practice of popular religion on the mountain—including shamanism—has been banned.

Today there are three large, thriving Buddhist temples on the mountain: Kapsa, Sinwŏnsa, and Tonghaksa. Each of these temples has its own monasteries, situated in the various valleys that the temples occupy: four in the Kapsa Valley; six in the Tonghaksa Valley; and six in the Sinwŏnsa Valley. In terms of physical size, Kapsa is the largest of the temples, Tonghaksa the second largest, and Sinwŏnsa the smallest. Eighteen other smaller, independent temples are scattered across the mountain. All of these have been built since the 1940s, and half of them are situated in Tonghaksa Valley.

According to a series of traveler's documents, Kapsa has always been the largest temple, garnering royal support during the Chosŏn period (1392–1910). In the late Chosŏn period, by contrast, Sinwŏnsa and Tonghaksa gradually declined. An eighteenth-century traveler (Yi Hae 1712) recounts that there were several hundred monks living at Kapsa, several times the number of monks at Sinwŏnsa. Another document mentions that no more than six or seven monks remained at Tonghaksa (Nam 1731). Since an earlier document (Song Sang-gi 1700) described Tonghaksa as a famous temple, it seems that Tonghaksa declined rapidly within a very short period of time. A later traveler's account (Kwŏn Kam 1793) confirms that Sinwŏnsa had been completely destroyed, leaving only ruins. Based on this early documentation (Nam 1731), one can surmise that Sinwŏnsa experienced this rapid decline within a period of less than sixty years (1731–1790).

After the Silla Kingdom (57 BCE–935 CE) conquered the Paekche Kingdom, the monk Ŭisang appointed Kapsa as one of the ten temples representing the Hwawŏm (Flower Garland) Sect. The temple is situated in a valley from which one can look up to the peak Ssalgyebong. All of the buildings in the temple complex were burned down in 1597 during the Imjin War (1592–1598), but they were rebuilt from 1604 to 1654 (Chŏng and Sŏ 1996, 77). The buildings were reconstructed once again starting in the late eighteenth century and finishing in the late nineteenth century: P'yoch'ungwŏn (1783), Chŏkmuktang (1797), and Sungmukdang (1899).

Tonghaksa, by way of contrast, was originally established in 724 during the Silla Kingdom. At the time, it was called Ch'ŏngnyangsa. The founder of the Koryŏ dynasty, King T'aejo (reigned 918–943), decreed in 920 that the temple would be the site where prayers for national prosperity were to be said. In 936, a Confucian shrine, Tonghaksa, was built at the temple to house rituals for Pakhyŏkkŏse, the founder of the Silla Kingdom, and for Pak Chesang, a loyal subject of the Silla Kingdom who died in Japan after rescuing the nineteenth king's younger brother from Japan (*Chohon'gakchi* 2004).

In 1394, during the rein of King T'aejo, a Confucian altar was established to commemorate Chŏng Mong-ju (1337–1392), a loyal subject from the Koryŏ dynasty. In 1400, this altar was developed into a Confucian shrine named Samŭn'gak. Two more loyal subjects, Yi Saek (1328–1396) and Kil Che (1353–1419), were enshrined there. In 1924, the mortuary tablets of two more loyal subjects were also added to the altar to encourage a national ethos during the Japanese occupation. During the rein of King Sejo (1456–1458), a Confucian shrine, Ch'ohon'gak, was also built to comfort the spirits of those killed in the political struggle against King Sejo. The shrine was completely destroyed by fire in 1728 and not reconstructed until 1869–1883. In 1903, twenty years after the reconstruction, the royal court bestowed on the shrine a name plaque: Sungmojŏn. The Sungmohoe Association was organized in 1963 to perform Confucian rituals twice a year, in spring and autumn, at the temple. In 1992, another hall, Sungmoje, was constructed to host meetings commemorating past loyalty.

It has been suggested that Tonghaksa was transformed into a Confucian academy, Tonghak Sŏwŏn, sometime before the rein of King Chŏngjo (1777–1800) (Han Wu-gŭn 1994, 21). The academy continued to function as an institution until it recovered its former status of temple in 1836. Even today, the Confucian gate, Hongsalmun, instead of the Buddhist gate, Ilchumun, stands at the entrance to the temple. To the left of the gate, a pavilion named Serokchŏng stands beside a stream, with a spatial organization implying that by this time the social status of neo-Confucianism on Kyeryong Mountain was considered superior to that of Buddhism and shamanism.

Sacred Sites in Shamanism

Shamans *(mansin)* and their clients make pilgrimages deep into the mountain to supplicate the mountain gods and other spirits and deities. Shamans by themselves sometimes camp out near peaks or in remote ravines famous for their shamanic and geomantic powers. Here, they bow and pray at shrines, in part to enhance their own supernatural powers. Often, shamanic rituals *(kut)* are performed on hilltops or high peaks, in deep mountain ravines near waterfalls, or up against the cliffs that are perched on dramatic slopes (Mason 1999a, 143).

Kyeryong Mountain is perhaps the most important of all the mountains that shamans visit. On the mountain, there are many sites that entice the shamans and their followers because of their mystical appearance and their secluded location. Of these sites, the most popular are two peaks and two ponds: Sambul Peak (Three Buddha Peak), Yŏnch'ŏn Peak (Connecting Sky Peak), Sutyong Pond (Male Dragon Pond), and Amyong Pond (Female Dragon Pond). Sambul Peak is one of the most sacred sites on Kyeryong Mountain because it symbolizes several of the important deities in Korean shamanism. Many women have come either with shamans or by themselves to pray for the birth of a son beneath this peak. Neo-Confucian principles in the Chosŏn period demanded that women give birth to sons in order to carry on the family line. Consequently, women in desperate need of a son turned to shamanism to help them resolve this problem. The extraordinary emphasis on agnatic kinship (and primogeniture) situated these shaman performances in an unlikely yet important relationship to the neo-Confucian power structures. In this way, shamanism could coexist with neo-Confucianism. The birth deities—*samsin* (three spirits) or *sambul* (three Buddhas)—were the focus of these *kut* (Covell 1986, 86–88). Together these deities representing procreation became a principal symbol of fertility in Korean shamanism. Even during the Chosŏn period (1392–1910), neo-Confucianism had no remedies for the anxieties caused by childlessness, childhood illnesses, and child rearing. Even though the proper performance of ancestor worship was thought to ensure the prosperity of descendants, it could not satisfy the emotional needs of women (Walraven 1999, 185). Accordingly, shamanism was able to occupy an important spiritual—and spatial—niche, even during a time when the religion and its practitioners were persecuted.

Since both Taoism and Buddhism influenced shamanism, it is not surprising that Taoist and Buddhist icons exist within shamanic practice. Indeed, there is no purely unadulterated shamanic representation of deities. The *sambul*, for example, are a meld of Buddhist representations coupled to a shamanic concept (Covell 1986, 86). While the shamanic concept of *samsin* has a long history, during its development it acquired distinctive Buddhist elements. Indeed, the *samsin* are almost always depicted as three monks in religious robes and are therefore often called *sambul*, or the "three Buddhas."

Consequently, even during a time of Buddhist and Confucian hegemony, shamanism—disguised in Buddhist clothes—could still help women address their desperate need for sons. Shamanism was able to survive during the Koryŏ period (918–1392) and the Chosŏn period (1392–1910) only by catering first to Buddhist needs and then to Confucian needs. An illustration of this tactic of survival being played out in the landscape of the mountain can be seen at the Buddhist monastery, Kyemyŏng Chŏngsa. Situated under Sambul Peak, the temple belonged to Tonghaksa. Although it is now prohibited from hosting shamanic activities, it had once served as a shamanistic shrine. Its strong link to shamanism was interrupted by Tonghaksa's claim to the area and Tonghaksa's subsequent push to build a temple beside the archaeological site of Ch'ŏngnyangsa in 1997 (Yi Kil-gu 1999, 211–213). Prior to that project, Tonghaksa had allowed Kyemyŏng Chŏngsa to be used as a space where shamanism could thrive, albeit clad in Buddhist clothes.

Yŏnch'ŏn Peak (Yŏnch'ŏnbong) is another sacred site where many shamanists come in search of spirits. From here one can look farther up to the nearby Ch'ŏnhwang Peak (The Heavenly Emperor Peak), the highest peak on the mountain. That is where the heavenly spirit descends. Just below Yŏnch'ŏn Peak lies the Buddhist monastery Dŭngwunam. This small monastery was established by Sinwŏnsa, and now not only Buddhists but also shamanists visit it (Yi Kil-gu 1999, 261). Shaman altars or shrines are a frequent feature of the landscape along the route from this monastery to Yŏnch'ŏn Peak. Kŭmnyongam is another monastery owned by Sinwŏnsa, and it too allows shamanists to stay there for their prayers and rituals. Thus, like Kyemyŏng Chŏnga, these two monasteries also provide a space for shamanism in places otherwise encoded as Buddhist.

Two sexually charged ponds—Sutyongch'u (Male-Dragon Pond) and Amyongch'u (Female-Dragon Pond)—are also sacred sites for shamanists on the mountain. The names of these ponds allegedly derive from a myth: At one time, a male dragon and a female dragon resided separately in these ponds, but they would swim underground to meet with each other. Finally the dragons ascended back into the sky from whence they had come. The ponds are believed to be the best sites for meeting the three spirits at the same time, since heaven, land, and water all converge there.

Despite the mystical sonic orchestration of water crashing over rocks and the seemingly mysterious shade provided by the thick trees, the deep ponds are little more than potholes, approximately 4 to 5 m deep. They were formed in the rocky valleys by erosion from the nearby waterfalls. The division of the ponds according to gender may derive from their overall shape: From the air, Sutyong Pond resembles a penis, while Amyong Pond resembles a vagina (Fig. 7.1). For this reason, women desperate for the birth of a son would come to Sutyong Pond, while men in need of a daughter went to Amyong Pond. Interestingly—and separate from the use of the ponds as the sites of shamanic worship—during the Japanese occupation, twelve Confucian literati who opposed colonial rule held regular meetings at Amyong Pond (Yi Kil-gu 1996, 225), eventually inscribing their names on the rocks behind the sacred pond as a physical manifestation of their oath not to become Japanese subjects. It is clear that the inscription itself on the rocks of a sacred spot on a mountain of extraordinary historical and spiritual significance for Korea was meant to demonstrate their determination to resist the Japanese.

Fig. 7.1. Sutyongch'u, or the Male Dragon Pond.

The Sacred Space in Popular Religion

In the late Chosŏn period, there were only a few houses in the Sindoan Valley, with very little arable land (*Silhwa Kyeryongsan* 1965). In 1918, about a hundred households who believed in the prophecies of the *Chŏnggamnok* moved to the valley. In 1924, another thousand or so believers in Sichŏn'gyo, a sect of Tonghak, also settled in the valley. Most of them were from the provinces of Hwanghaedo and P'yŏngando, as a prophecy in the *Chŏnggamnok* indicated that no trace of human life would be seen in those provinces for three years (Yi Min-su 1993, 17).[1] Another geomantic prophecy said that the Sindoan Valley would become the center of the world in the new millennium, and this also helped lure them there. The second leader of Tonghak interpreted the book of *Chŏnggamnok* differently, predicting that this world would come to an end and would be followed by a new world in 1924 (*Silhwa Kyeryongsan* 1965). The result of these prophecies was that a large village of nearly 5,000 households (7,000 persons) sprang up in the valley in 1924. After 1924 passed without any dire events, a second prediction proposed that this world would come to an end in 1954. Those who believed in this prophecy continued to join with those who had already settled down in the valley. Most of them were from groups other than Sichŏn'gyo; by the early 1950s, these millenarian sects numbered more than two hundred.

These religious groups were all basically similar, since they held as a fundamental tenet that the Sindoan Valley would become the center of the world in the new millennium (Tonga Ilbo 1923–1924). For all of the groups, the *Chŏnggamnok* predicted that the Chosŏn dynasty would come to an end and the Chŏng dynasty would replace it, with its capital in the valley. The followers dreamed that the valley would be the place from which a bachelor messiah, Chŏng Toryŏng, would govern the new world. Furthermore, they believed that sixty-six countries would come to the valley and pay tribute to this new Chŏng dynasty. That King T'aejo (reigned 1392–1398) had planned to construct the royal capital in the valley lent credence to this dream. Not too surprisingly, the *Chŏnggamnok* experienced significant popularity in both the late Chosŏn dynasty, during the Japanese occupation, and also during the Korean War.

These groups also shared the geomantic idea that the valley was situated at a source of considerable power due to the relationship between the mountain energy and the water energy. Geomancers believed that the power of the mountain originated at Mt. Paekdu and flowed in a circle through Mt. T'aebaek, Mt. Sobaek, Mt. Chiri, Mt. Tŏgyu, and Mt. Taedun, arriving finally at Mt. Kyeryong (Fig. 7.2). In contrast, the flow of the water formed two patterns, one small and one large. In the small pattern, water flowed in a circle from Mt. Kyeryong through the Sindoan Valley into the Kŭm River. From there it made a circle into the Yellow Sea. In the large pattern, water flowed from the origin through the Kŭm River to form a circle around Mt. Kyeryong and back into the Yellow Sea. The geomancers imagined that the overall patterns of energy flow on the mountain and in the water composed the visual image of *taiji* (in Chinese, literally two gigantic opposite powers). The Sindoan Valley was located at the center of the flow patterns and stood as the source of those opposite powers: *yin* and *yang* (see Fig. 7.2). Because of this extraordinary positioning—essentially as the source of all power in the universe—it was believed that the valley would become the political center of the world in the new millennium.

The Sindoan Valley was not only a source of remarkable power, but geomancers also considered it the most auspicious of all places, surrounded as it was by mountains in four directions, symbolizing the imaginary animals blue dragon, white tiger, black turtle, and red peacock. The mountains and hills surrounding the valley in a circle also resembled a golden hen holding her eggs either at her breast or under her wings (Yun Sŭng-yong 1994, 590–592). Another interpretation proposed that the arrangement of mountains and hills around the valley resembled a dragon playing with a pearl. In either case, the residents of the valley believed that such topography would bring them fortune.

This short survey reveals that places such as the Sindoan valley were constructed and subsequently experienced as the focus of these beliefs. *Chŏnggamnok* first and then Tonghak offered myths and ancillary beliefs to

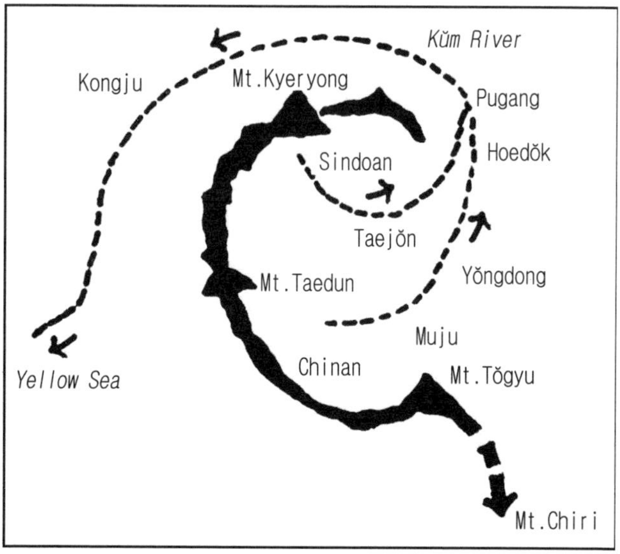

FIG. 7.2. The overall pattern of flow in the mountain and water around Sindoan.

individuals who were faced with dissolving boundaries and shifting frontiers, particularly at times of heightened social and political uncertainty. Consequently, the places became filled with symbolic and representational meanings. All of the belief systems that found shelter in the Sindoan valley reflected profound syncretic tendencies in their doctrines and pursued the integration of religious elements of the religious traditions from which they emerged, as well as from Confucianism, Taoism, Buddhism, and shamanism. The culture of the valley was one that allowed for—even insisted upon—multiple understandings. The powerful concept of entanglement can perhaps best be used to explain these syncretic tendencies. The familiar Taoist image of the *yin* and *yang* symbol helps us visualize claims about the entanglement of domination and resistance (Sharp 2000, 20). Although the symbol implies a dynamic balance of opposed forces, it also implies that there is to be no complete separation between two seemingly opposed practices, in that one will always contain at least the seed of the other. In both China and Korea, there is a preference for blurring, not sharpening, religious identity, expressed in the ideal of "the three religions being one" (Buswell 1999, 141–148). The social preservation of religion as a major institution within secular societies has been won in part through the successful creation, protection, and nurturing of symbolic places (Harvey 1993, 23). It was no different in this valley.

The Control of Kyeryong Mountain by the State

In 1975, the Korean government decreed that Kyeryong Mountain was to be preserved as a national park and that all the followers of the various popular religions who called the mountain home, including the shamanists, should leave. Some of those who resisted the order were arrested and thrown in jail. Within two years, many who had reluctantly followed the directive returned again. In 1979, the state launched a second initiative to clear these groups from the mountain. Negotiations with the groups this time included financial subsidies or land grants to those who voluntarily left.

In 1983, the state decided to establish in the Sindoan Valley a military command center for land, sea, and air forces. Subsequently, in order to protect the military base, the state forced all of the adherents of the various popular religions to leave the valley. Some suggest that the buildings of the headquarters were intentionally located in the most geomantically auspicious spots in the valley. The main building, in particular, was built in the shape of an octagon, symbolizing the "Eight Trigrams" in the *Book of Changes* (Yi Kil-gu 1999, 167). The location of the headquarters is also believed to be surrounded by the imaginary animals: a dragon ascending the sky, a hen holding its eggs, a tiger lying flat on the ground, and a serpent coiled on the ground (Fig. 7.3). These imaginary animals are of course swift and strong, quite able to protect the headquarters from any outside attack.

While the construction of the military headquarters was completed in 1989, the planned city, Kyeryongsi (Kyeryong City), was still under construction. The city was designed to accommodate the military and their family members. The original plan was to build a "bed city" that could support a population of 150,000 by the year 2011 (Kyeryongsi 2004). The current population is now slightly over 30,000, of which military family members comprise 60 percent. The city hall also planned to develop the urban area in a manner that would conform to geomantic interpretations of the site, insisting that the layout of the city when completed should look like "a golden hen" flying up to the highest peak on Kyeryong Mountain, Ch'ŏnhwangbong.

Urban planners went on to imagine the physical parts of the hen: comb, eye, neck, breast, heart, body, and tail and matched the use of each planned district with each of these physical parts: the administrative district with the comb; the commercial district with the neck and breast; the cultural district with the eye; the old city, highway, and freeway with the body and tail. The heart, a hill called Hyojisan, was to remain as a public park with trees

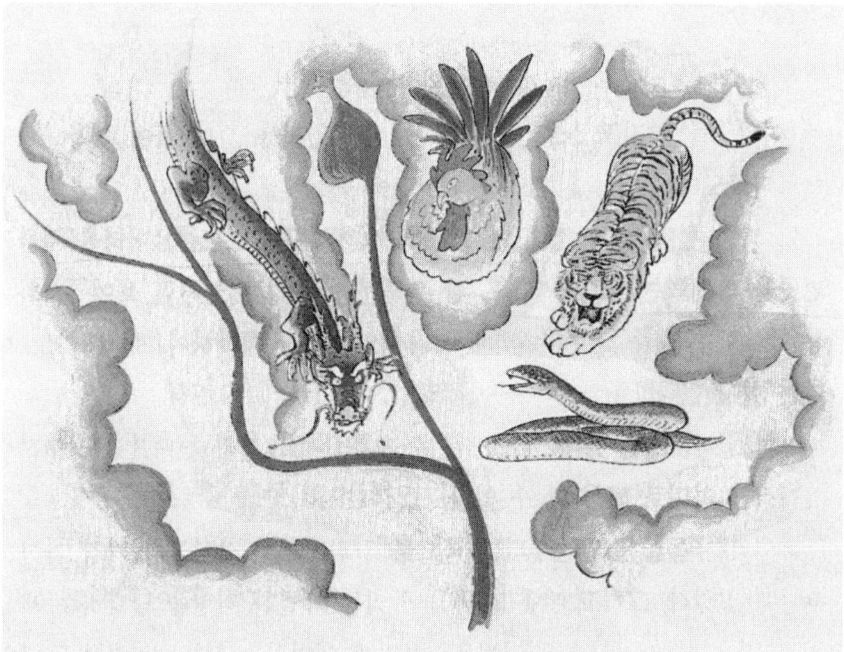

FIG. 7.3. The geomantic interpretation of the Sindoan Valley.

because it was not to be harmed. The result was that the administrative, commercial, and business districts would be located in the most geomantically auspicious spots.

Harvey points out that the fierce contest over images and counterimages of places is an arena in which the cultural politics of places, the political economy of development, and the accumulation of a sense of social power exist (1993, 23). In the urban planning of Kyeryong City, the political-economic possibilities of place construction were colored by the evaluative manner of place reconstruction. By the same token, the creation of symbolic places is not a given. Rather it is nurtured in a painstaking fashion and fought over, precisely because of the hold that place can have on the imagination.

The Geography of Resistance or Entanglement

Shamanists like to pray for spirits at night, perhaps because of the mysterious atmosphere that pervades the mountain after dark. Even before 1975, Buddhist halls or monasteries scattered over the mountain housed shamanists who would often spend the night. Despite their Buddhist appearance, these halls and monasteries were in fact shaman shrines *(kuttang)* where only the shamans performed. To ward off the suspicious eyes of the government, the shamanists pretended to be Buddhists.

Before the official ban on shamanism in 1975, more than a hundred shaman shrines dotted the mountain (Fig. 7.4). In general, these shrines were very small and individually owned. The owners were predominantly women who wanted to pray for their own spiritual power or health. Prior to 1975, the mountain peaks and ponds were crucial to the location of the shrines. After 1975, by contrast, shaman shrines near mountain peaks and ponds all but disappeared; those near Buddhist temples managed to survive, albeit in limited locations (Fig. 7.5). This shift in the location of the shrines suggests that power relations can produce discontinuous spaces. Those resisting the dominant power are likely to transgress or move between these newly formed, discontinuous spaces (Pile 1997, 14).

From the 1990s, when the state's control of the mountain weakened, shaman shrines no longer needed to masquerade as Buddhist halls or monasteries. Consequently there has been a significant increase in the number of these shrines. These new shrines usually fly white and red banners high in the sky and place signs outside. When a shrine is large, it often consists of several halls and open spaces where *kut* can be performed. One such shrine began to develop in the late 1970s when the Saemaŭl Undong (New Village Movement) was launched (Ku 2001, 37). As shamanic rituals were officially prohibited in the people's houses, shamans began to consider open spaces near Kyeryong Mountain as suitable for their rituals. Many of these shaman

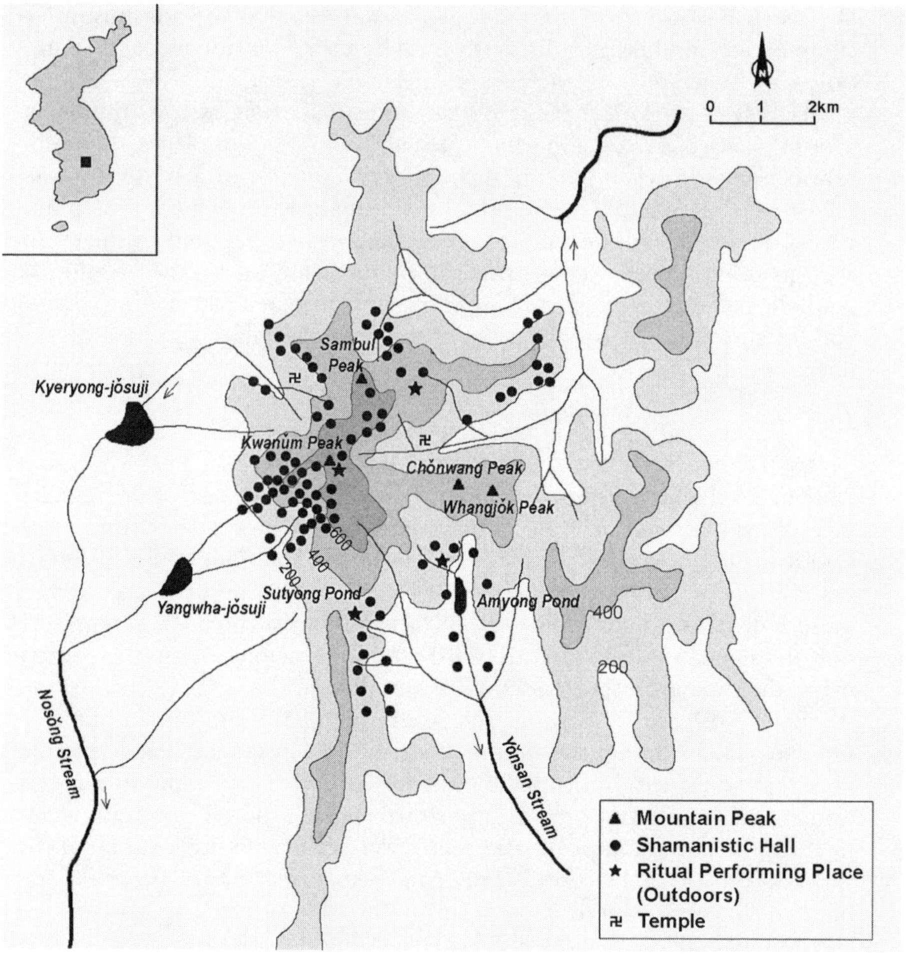

FIG. 7.4. The distribution of shamanistic halls on Kyeryong Mountain before 1975.

shrines are now located in the foothills of Kyeryong Mountain, outside the perimeter of the national park.

After 1975, the shamanists who could not survive the pressures of state control relocated their shrines to other sites from where they could see the sacred peaks of Kyeryong Mountain. The Sangsilli Valley and the entrance to Tonghaksa appealed to them the most. Shrines at the entrance to Tonghaksa have thrived, partly owing to easy access from Taejŏnsi (Taejŏn City). Similarly, the shrines in the Sangsilli Valley have been successful in attracting clients, as they are located on sites from which Sambul Peak can be clearly seen. In this valley, there are also two small Buddhist temples that

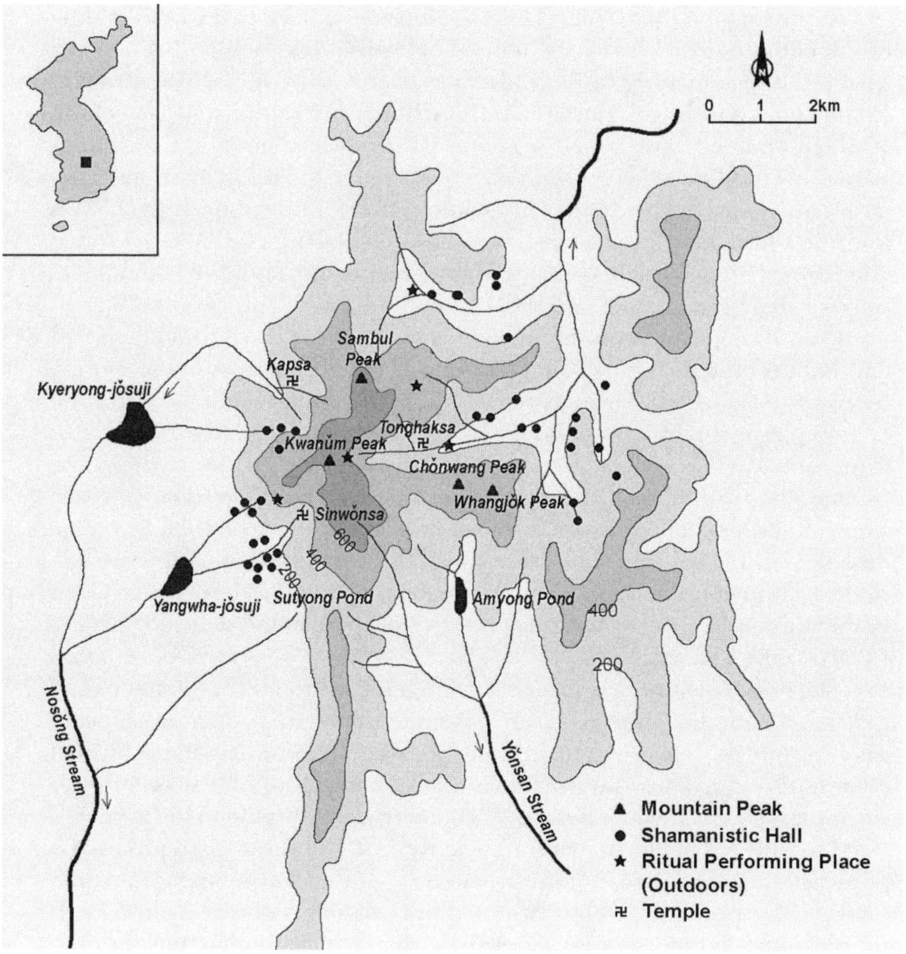

FIG. 7.5. The distribution of shamanistic halls on Kyeryong Mountain after 1975.

also have shaman shrines: Chŏngnyongsa and Chabisa. The former moved from its prior location under Sambul Peak to the valley, and the latter took over a shaman shrine named Yŏmburam. This suggests that many shamanists who had to leave sites within the boundaries of the national park moved their shrines to the Sangsilli Valley and outside of the national park. In this manner, the state authority unwittingly produced new spaces through the process of declaring boundaries and controlling movement within and across those boundaries. The resistance moved outside these boundaries and refused to be confined to the newly established authorized spaces of domination (Pile 1997, 3).

Kŭmnyongam, or Golden Dragon Monastery, is located in the middle of the climbing route from the temple Sinwŏnsa to Yŏnch'ŏn Peak. It is a small temple consisting of Buddhist and shamanic buildings: Taewungjŏn (Main Hall), Sansin'gak (Mountain Deity Hall), and Kŭmnyong Tongch'ŏn (Golden Dragon Cavern). Taewungjŏn is a Buddhist building that houses statues of Buddha, while Sansin'gak is a shamanic building that houses the mountain deities (Grandfather or Grandmother). Kŭmnyong Tongch'ŏn is a shrine where the deity of the Dragon-King is worshipped. A valley called Yonggung (Dragon's Palace), with some seventeen sites where shamanists pray for their own wishes, stretches out behind this shrine.

Even though the monastery belongs to Sinwŏnsa, it is privately owned and built on the former site of a shaman shrine. The owner of the monastery kept the shaman shrine's name—Kŭmnyongam—while erecting the contemporary buildings on land rented from Sinwŏnsa (Ku 2001, 226). Consequently it is only at this temple that shamans can pray within the perimeter of Sinwŏnsa. The monastery has been a contradictory space where Buddhism allows shamanism to flourish in private, while in public driving it out. The opposite powers of Buddhism and shamanism are not necessarily in conflict. Rather, resistance in one place may be complicit with domination in another. This initiates the concept of entangled geographies (Sharp 2000, 24).

The ponds of Sutyongch'u and Amyongch'u are today completely closed to human entry for military reasons. Despite these circumstances, shamanists continue to steal their way into the ponds at night in search of spiritual communication. These sacred sites are far too important for shamanists to stop coming and praying. Before 1983, many women came to Sutyongch'u Pond to pray for the birth of a son. Inscribed on the rocks surrounding the pond are the many names of those who have prayed there. At night, in particular, these ponds were busy with women or shamans who dedicated foods to spirits and lighted candles. After the public ban on human entry, only a small number of shamanists with strong spiritual needs dared to visit. These nocturnal visits should be seen as acts of resistance, acts that often take place under the very noses of the oppressors; both resistors and oppressors, of course, offer their own distinct spatialities (Pile 1997, 2). In this context, the shamanists engage the logic of nomadism, following their own hidden tracks in a field outside of their jurisdiction.

The heaviest concentrations of shaman shrines has always been at sites from where Yŏnch'ŏn Peak can be seen. For instance, many of the shaman altars or shrines near Dŭngwunam were established after the official ban on shamanism (1975) and the construction of the military headquarters (1983). Indeed, many of the shaman shrines around Sutyong and Amyong Ponds ultimately ended up with moving there. At Dŭngwunam, simple altars are built at the base of rocks that form a high ridge on the side of a slope. Usually this is an overhang that forms a hollow too shallow to be a cave, and it is here that the shamans situate their shrines. One often sees two or three white candles burning and perhaps a bowl of fresh water at these small shrines or altars, and chanted prayers are offered there. The altars at the shrines are more often than not simply a flat natural or hewn stone; these are called Sansindan, the altar for the mountain deity (Mason 1999a, 40).

Chinese letters discovered on a nearby cliff have made Yŏnch'ŏn Peak even more sacred. The adherents of *Chŏnggamnok* apparently inscribed eight Chinese letters as ciphers, indicating that the Chosŏn dynasty would come to an end in 1910. The inscription represents the popular dream at the time that the troubled world would soon be a thing of the past. To add further weight to this notion, the name "Yŏnch'ŏnbong" means "a peak to be connected to the heaven." In one interpretation of this name, the spirit would come down to the highest summit, Ch'ŏnhwang Peak. On this summit one still finds two altars: Chŏndan (Altar for the Heavenly Spirit) and Sanjedan (Altar for the Mountain Spirit). The presence of these altars on Ch'ŏnhwang Peak has certainly contributed to the sacredness of Yŏnch'ŏn Peak; there one can have access to both the heavenly and mountain spirits. It is not known when Chŏndan was first built by the followers of the Sichŏn'gyo sect of Tonghak. Sanjedan, on the other hand, is known to have been erected in 1944 by the fourth head of Tonghak. After they moved to the Sindoan valley, these Tonghak devotees began to dedicate annual rituals on these altars on April 1 (lunar calendar) to the heavenly spirit Sangje, the mountain spirit Kyeryong Sansin, and Ch'oe Che-wu, the founder of Tonghak (Yi Kil-gu 1996, 289).

Sometime after 1945, the American military, in cooperation with the Korean military, established a relay communications base on the summit of Ch'ŏnhwang Peak. In 1981, a civil telecommunications company, Han'guk T'ongsin, took over the base and erected two large towers on and below the summit. Citizens from Taejŏn City protested, asking that the iron tower on the summit be removed because it weakened the telluric energy *(chigi)* on the mountain. The company eventually agreed to relocate the tower below the summit. The summit was opened to the public in 1998. Subsequently, Tonghak believers tried to appropriate the altars on the summit, intending to use them to perform their annual rituals and thereby strengthen their group identity. Engagements in the politics of location such as these clearly involve the definition of boundaries that are not seen as fixed, impermeable, or permanent (Pile 1997, 26). In the struggle to define an alternative way of living, Tonghak followers attempted to occupy the summit because of its strategic location.

During the Chosŏn dynasty, neo-Confucianists were radical idealists whose goal was to replace popular customs with neo-Confucian culture.

Inevitably, their attempts to change long-standing practices and ancient beliefs met with resistance. Faced with resistance from the popular level, they had to enter into compromise with the followers of popular beliefs as a means to impose neo-Confucian elements onto these popular rituals (Walraven 1999, 172–173). When the irregularities of some cults were too glaring, government officials preferred to accommodate them in some way. Neo-Confucianists could not ignore the old tradition of royal support for the worship of mountain deities once they gained national power (Mason 1999, 157). Out of all the shrines dedicated to the mountain deity, however, only one remains today in its old form within the perimeter of Sinwŏnsa on Kyeryong Mountain.

This is the shrine Chungaktan (formerly Kyeryongdan).[2] The Japanese colonial government stopped the performance of all rituals at this altar, but in 1998 local cultural leaders revived the tradition of grand ceremonies that combined the three elements of neo-Confucianism, Buddhism, and shamanism. Indeed, with its tolerance for indigenous cultures, Buddhism has acted as an intermediary between the two opposing powers of neo-Confucianism and shamanism throughout history. Today the local government of Kongju City officially supports the ritual dedicated to the deity of Kyeryong Mountain at the altar Chungaktan (Central Mountain Altar). This latest accommodation is clearly a reflection of the sociopolitical processes of the state. These processes are ones of perpetual movement, negotiation, changing alliances and affinities, co-options and infiltrations, always contingent upon the particularity of spatiotemporal conditions (Sharp 2000, 11).

Conclusion

These geographies of resistance, involving in large part shamanism, are scattered across Kyeryong Mountain, highlighting the discontinuities of territorialization that characterize this space. These geographies are most easily identified near the most sacred sites, such as Sambul Peak, Amyong Pond, and Sutyong Pond.[3] At many of these sites, the solution to the later official suppression of shamanism was to resist by melding Buddhist imagery *(sambul)* with shamanic concepts *(samsin)*. The two ponds, Amyongch'u and Sutyongch'u—both sacred shamanic sites—were never relinquished, despite the military's attempts at exclusion. Even after the construction of the military headquarters (1983–1989), shamanists still managed to get to these sacred sites to perform rituals and pray, thus contesting the official use of these sites as nonsacred and in the service of secular goals.

The entanglement of shamanism with Buddhism in various spiritual and temporal patterns contributed to the survival of shamanism on the mountain, even when it had to exist as a subordinate power. If one overlooks the position of Buddhism as an intermediary in the relation between the state and shamanism, it is difficult to explain the survival of shamanism on the mountain. Indeed, such entanglements of power can be easily found in the Sinwŏnsa Valley, where Buddhist halls were the most tolerant of the practice of shamanism. The geographies of entanglement, in various spatial formations, can also be seen along the perimeters of Kŭmnyongam monastery. Here, a variety of shamanic practices are always permitted, even during the daytime.

Sindoan Valley is the one place on the mountain where conflicts over the use of space have been the most apparent. In the valley, the geographies of entangled power relations have intersected with the messy and inherently spatialized entanglements of domination and resistance. Prior to the state intervention in 1975, there was serious competition over space in the valley among the adherents of various popular religions. They all shared beliefs in the geomantic prophesies of the *Chŏnggamnok,* indicating that the valley would become the center of the world in the new millennium. In order to dominate in the competition over sites and spaces, each of them created their own vision of paradise even while sharing the same geomantic prophesies. The result was a geography of domination and resistance as a contingent and continuous bundle of relations. These entanglements consequently enacted a contested encounter within and between dominant and resistant practices—practices that were necessarily hybrid rather than simply binary.

After the construction of the military headquarters in the valley, the state actively appropriated spaces from the various groups of believers. In planning Kyeryong City, the state reinterpreted the traditional geomantic idea of the valley to create its own place myth, thereby revealing that a geography of domination often comprises a web of tension across space, hybrid assemblages of knowledge, vocabularies, and judgments. In other words, there has been a fierce contest over images and counterimages of sites and spaces in the Sindoan Valley. In this case, the political economy of development involved cultural politics of places and the accumulation of social power. The state has tried to dominate the valley as a power by not only appropriating the material spaces but also creating symbolic places. This study of Kyeryong Mountain has made it clear that mountains in Korea can be seen as complex spatial texts that need to be deciphered. Geographies of dominance and resistance or entanglement have complicated the meanings of sites and spaces throughout time.

Notes

1. The name of this movement changed from Sichŏn'gyo to Sangjegyo immediately after their move to the valley. The name of the movement was subsequently changed to Ch'ŏnjin'gyo in the 1950s. Depending on the main aspects

of doctrine, the religious groups in the Sindoan Valley could be divided into five categories: Tonghak, Buddhism, Tan'gun, Christianity, and shamanism. The sect, called Sichŏn'gyo, from Tonghak, was the majority in terms of resident number, while Buddhism was the largest in terms of sectarian number (*Silhwa Kyeryongsan* 1965). Each congregation from Buddhism is in reality an independent sect separate from Buddhism, even if it took its architecture style and name from the Buddhist temple.

2. It was called Kyeryongdan until the reign of King Kojong (1863–1907), when its name was changed to Chungaktan (Central Peak Altar).

3. The name of the peak (Sambulbong) originally came from the shamanistic concept of *samsin,* the deity of procreation or pregnancy. Many women used to come here with shamans or by themselves to pray for the birth of a son.

8

Kyŏngju Namsan

Heterotopia, Place-Agency, and Historiographic Leverage

Robert Oppenheim

Consider the most subtle and most dangerous orator of Greece preparing, before the assembled Trojans, to weave the glittering web of his words.

—Detienne and Vernant, *Cunning Intelligence in Greek Culture and Society*

If you don't go up Namsan, you can't say that you've seen Kyŏngju.

—Kyŏngju saying

Namsan means, simply enough, "South Mountain," and there are Namsans south of several cities in Korea. The Namsan of Kyŏngju—a city that as the site of the ancient Silla capital has more recently been a magnet for archaeology and tourism—has two main peaks with heights of 468 and 494 m and spans roughly 8 km along its north-south axis and 4 from east to west. The terrain is craggy and mostly wooded, scenic with gnarled pines, although sunlight breaks through in frequent clearings, and from some spots one can see a long way. On Namsan, several functioning Buddhist temples, including one under reconstruction, an observatory pavilion at an especially advantageous point, and even a sort of mountain inn serving vegetarian "wild" cuisine can all be found. The main basis of Namsan's importance, however, is its numerous historic relics and cultural objects: royal mound tombs, stone pagodas *(t'ap),* and Buddhas and bodhisattvas, alone or in triads, carved of stone and free-standing or incised into living rock, mostly dating from preunification Silla. Like others of that era, before the apex of the Silla state and Silla royal power, Namsan's Buddhist figures are mostly simple in their execution and human in their scale. Usually, they smile. When one hikes Namsan, thirty seconds or thirty minutes, another turn or another twenty can bring a Buddha from around a corner. Knowing where to go is yet another matter: Namsan's paths are many and varied in their marking and clarity. They twist and divide. While some routes are well traveled and some monuments easily found, to search for a particular relic without a guide can be an invitation to hours of frustration.

The users of Namsan are various and include tourists, alpine clubs, academics, and shamans on mountain pilgrimages (Kendall 1985, 128–131), to name but a few. In this chapter I am principally concerned with the Namsan of certain Kyŏngju native-place *(hyangt'o)* historians and locally centered lay Buddhist organizations of an interlinked group of actors that form a segment of what in newspapers and the like is sometimes identified as Kyŏngju's "culture world" *(munhwagye)*. Through study and ambulatory *tapsa* (field investigations of historic objects), they have contributed greatly to Namsan's contemporary fame and to the sense of its obligatoriness, embodied in my second epigraph above, for any who would presume to know the ancient city.

In a broad sense, this essay thus responds to calls to chronicle Korean historiographies that are neither simply official nor oppositional (e.g., *minjung*) and that are also not the product of professional historians (de Ceuster 2001; Delissen 2001; Walraven 2001; cf. Trouillot 1995, 19; Wells 2003). It examines local cultural texts, practices, and landscapes in a post-1987 (and 1992 and 1995) era of normative "localization" *(chibanghwa)* (cf. Kim Kwang-ŏk 1994a). Namsan is, certainly, a heterotopic site (Foucault 1986): an alternative space and a space of alternatives. Yet since Kyŏngju is a national showpiece, where even for "locals" the question of "who owns the past?" has rarely had available a plausible, simply "local" answer, Namsan is also more: a site of leverage. When writing about and equipmentally with Namsan, Kyŏngju amateur historians have been able to engage academic historiography in a relationship of relative equality, even preeminence; Namsan is one of the few topics on which academic historians cite Kyŏngju local historiography as authoritative (cf. Kim Sang-hyŏn 1995, 306–307). My story is thus not ultimately of incommensurabilities, of heterologic alternative topographies and historiographies in palimpsestic overlay (Halbwachs 1992), but rather concerns commensuration and its politics. How, I ask, does Namsan as a place serve to underwrite the authority of Kyŏngju local history and historians within a wider historiographic field? More fundamentally, how does place come to have agency?

In opening these questions, I seek to track between three contemporary literatures on place, objects, nature, agency, and power. From actor-network theory and its progeny, I draw a willingness to locate agency as distributed between human and nonhuman actors and a perspective on the complex spatiality of objects and places that are neither simply present *or* absent, proximate *or* distant. Entitivity-in-oscillation is precisely what allows for leverage (Callon and Law 2004; Latour 1983, 1996, 1999; Law 1999, 2002). Namsan works because it is at once, obligatorily, both a unique depth object of local culture and a token in the class "cultural object." From material culture studies and the anthropology of art, recently in revealing confluence with the anthropology of landscape (Bender 1993b; Hirsch and O'Hanlon 1995), I draw the notion of aesthetic and historiographical "traps" that "express the social efficacy of their creators and attempt to draw others into . . . exchange with those who have objectified themselves" (Miller 2000, 6; cf. Gell 1996). Namsan works because, as made, it is always ever "unfinished business" (Gell 1998, 80–81). Yet if many studies of landscape, art, and material culture alike manifest prior disciplinary commitments to conceptualization in terms of anthropogenic processes of subjective objectification (cf. Chapman 1998), a useful corrective can be found in environmental anthropologies and histories that trace looser and more multivalent interactions of the social and natural (Little 1999, 257–259; cf. Grove 1995; Greenough and Tsing 2003; Raffles 2002a). "Nature" might productively have a role not only in its own history (Hacking 2004, 16) but in the historiographical field. Shifting light, places to stand, foliage, all the material aspects of seeing, and not just "vision," have their politics. As a "glittering web," Namsan was constructed—assembled of parts (Hacking 1999, 49)—not just "of words" but of these as well.

"Look at the Mountain"

The rise of Namsan as a central and constitutive concern of Kyŏngju historiographic circles was conditioned by the marginality of the mountain to other projects of making history and place. During the 1970s, under the auspices of the centrally authored Kyŏngju Tourism Comprehensive Development Plan, the military Yusin state sought to remake the city at once for paying visitors and in its own self-image, which drew especially upon the martial valor of Silla kings and generals. The physical and ideological itineraries of this remaking skirted Namsan's edges and grasped the mountain as a neutral space, a container for various monuments, rather than a meaningful place in its own right (cf. Kyŏngju Kaebal Kyehoektan 1971, 57–58, 145–148). The oppositional *minjung* movement that grew to prominence in the 1980s, meanwhile, tended to delegitimate Silla's historical role and thus decentered Kyŏngju as a whole from its own historiographical practice in favor of other geographic touchstones (cf. Abelmann 1996, 11; Ch'oe Hyŏp 1994; Chŏng Kŭn-sik 1996; Yea 2000b). Scholarly attention finally had focused on Namsan in a colonial era survey of its Buddhist antiquities led by Oba Tsunekichi (Chosen Sotokufu [1941] 1994; Hwang Su-yŏng 1994), but after 1945 until the 1970s further study had been limited to extensions of the Japanese overview and specialized studies of individual monuments (Kim Sang-hyŏn 1995, 307; e.g., Chŏng Yŏng-ho 1967).

Inside Kyŏngju, the person most associated with Namsan is Yun Kyŏng-nyŏl, an author of several books on the mountain. In his eighties in 1997–1998 and long a self-proclaimed "Kyŏngju person," Yun was born in North Hamgyŏng Province, in present North Korea, and came south to stay during the Korean War, never to see his parents again.[1] After his arrival in 1956, Yun became one of the founding members of the Society of Friends of Silla Culture (Silla Munhwa Tonginhoe), a local amateur historical group initially made up largely of professional-class men (Kim Yun-gŭn 1995, 276).[2] Among its activities, since at least 1968 the Society of Friends has conducted formal monthly *tapsa* to historical sites in the region, including on numerous occasions to relics or temple sites on Namsan (Son Yul-lak 1987). Yun in particular was also heavily involved in the Museum School, a cultural education program for local children that the Friends, in cooperation with the Kyŏngju (National) Museum,[3] sponsored for many years (Yun Kyŏng-nyŏl 1994a).

According to his memoirs, by the mid-1970s Yun had ascended Namsan almost six hundred times, "visiting Buddhas to find the face of Korea," yet he had no intention of writing a book because, as a nonarchaeologist, he did not feel qualified. But as he walked Namsan time and time again, it began to trouble him that "foreign and Korean students" visiting and studying the relics of the mountain had only palm-sized maps with which to find their way. He also came to consider it "shameful" that, whereas "sixteen years after Japan occupied our country it put out a big book entitled *Buddhist Relics of Namsan,* in the thirty years since we [Koreans] regained our country there has not been one book on Namsan worth publishing." The dearth was so severe that the occasional newspaper or magazine article on Namsan had little choice but to quote the Japanese source directly. The sufficiency of this nationalist logic for restudying Namsan, however, was disturbed by the next resolution Yun recorded. "Gradually," he wrote, "the thought grew that, 'if somebody is going to write a book about Namsan, it should be a Kyŏngju person.'" The next day he began his task, working systematically, valley by valley, "changing even his way of walking" (Yun Kyŏng-nyŏl 1997, 215–216).

The question of the colonial study of Namsan and the issue of the situation of epistemology returned in Yun's account of his research in one valley—Kuksagol. Yun Kyŏng-nyŏl discussed his predecessor Koba Tsunekichi,[4] the principal researcher of *Buddhist Relics:*

> [Koba] was a part-time employee of the Chosen [colonial] government. At that time, if that sort of person appeared in Kyŏngju a *myŏn*[5] or village head would have guided him. These people have no reason to know mountains.

> The people of a village neighborhood[6] know mountains well. If after finishing an investigation he had gone into the *makkŏlli*[7] house at the end of the valley, there would have been woodcutters and old people from the village there. If he had offered them each a cup of *makkŏlli* and then asked them, they would have known a lot about the affairs of the valley. The owner of the *makkŏlli* house at the end of Kuksagol taught me a lot, and gave me a good deal of help (Yun Kyŏng-nyŏl 1997, 217).

In this rendition, Koba's study of Namsan was incomplete, structurally compromised by an epistemological gap that he failed to bridge. Yet this gap did not inhere in an essentialized relation between colonizer and colonized but rather in the relations that outsiders and political elites (Korean or Japanese) have, or fail to have, with those closer to the land. A nationalist logic—"only Koreans can know Korea"—was undermined by a localist incursion that reconfigured distance in social and spatial terms.

Yun's initial research on Namsan was completed in 1977 and published by the City of Kyŏngju in 1979 as *Kyŏngju Namsan kojŏk sullye* (Pilgrimage to the ancient relics of Kyŏngju's Namsan). Already a senior member of the Society of Friends, after that Yun also gained an independent, loose Kyŏngju following of persons interested in Namsan, many of whom consider themselves to be his students. Especially after his publication of subsequent books with larger press runs and wider distribution (e.g., 1993), Yun also became a minor celebrity nationwide. No doubt this was helped by his iconic appearance within a South Korea increasingly engaged in the commodification of culture and nostalgia, his oft-cited mane of white hair, and his preference for Korean dress (Yun Kyŏng-nyŏl 1997, 213; cf. Ruhlen 2003, 123–127). Indeed, his 1997 memoirs were published by Hakkojae, a Seoul art-historical publisher that counts cultural heavyweights like Yu Hong-june and archaeologist Kim Wŏl-lyong among its other authors. Yun has often been portrayed in these wider spheres as "the face of Kyŏngju culture," or, to quote the title of the 1997 book, the "Last Person of Silla" (Majimak Sillain). His fame helps connect the Kyŏngju culture world outward.

In a forward to his initial Namsan study, Yun (1979a: 5) declared the mountain "a treasure-house *[pogo]* of national culture *[minjok munhwa]*, weaving together history, art, and religion." He went on to recommend that "the wisdom of all our race *[kyŏre]* be assembled and an accurate and exact investigation be undertaken so that the true face of this sacred mountain can be revealed." Yun went on to describe the cultural monuments of Namsan, its Buddhist statues and carvings, stone pagodas, and royal tombs, in sections organized by the valley of their location. His 1979 book contains not only photographs but also "imaginative reconstructions"—impressionistic watercolor sketches of locations where "the original atmosphere cannot be photographed" because relics have been moved or have simply disappeared: Namsan not as it is, but as it might have been (1979a, 5, examples 179, 310–311). In an afterword, Yun proclaimed Namsan's inexhaustibility: "Not only are there many things to study, but there are various changes in each valley such that, no matter how many times one climbs it, Namsan is a mountain that never fails one" (312).

The main text began with a simple and dramatic heading: *"San Pwara,"* or "Look at the Mountain" (Yun Kyŏng-nyŏl 1979a, 31). It was soon made clear that the imperative was drawn from a legend of old Silla that Yun proceeded to recount. Yet, set off as the title is, it is also easy to read as imploring Kyŏngju residents and visitors to notice what is there right before their eyes. Indeed, later on the same page, Yun poses the rhetorical question that has come to define Namsan's location at the center of the local cultural imagination: "If a person who visits Kyŏngju does not see Namsan, can (s)he still say that (s)he has seen Kyŏngju?" Read literally, "Look at the Mountain" was an inaugural moment in the constitution of Namsan as a visual object.

What, then, did Yun see there? His early studies of Namsan established the mountain as a value-imbued element of the local landscape, a concretized metaphor for certain notions of appropriateness. The centerpoint of the Namsan aesthetic came to be "harmony" *(chohwa),* embodied in the relationship between cultural objects and their natural surroundings and manifest at various scalar levels.[8] In an article that appeared in a 1979 edition of the Friends' newsletter and journal *Ch'ŏn'go,* Yun's point of departure was one of the many stone pagodas found on Namsan. Korean art historians usually describe pagodas as having a certain, always odd number of stories—for Silla, three-storied pagodas are most common (Yun Kyŏng-nyŏl 1979b). However, this number refers only to the pagoda "body" *(t'apsin)*—its middle segment; pagodas also have "heads," including a finial of some sort, and bases made up of two additional dissimilar levels (Fig. 8.1). What distinguishes many of the pagodas on Namsan is that they are missing the lower base level; instead, the upper base level rests directly on the rock of the mountain. Yun contended that this design reveals the desire of the people of Silla to build a pagoda to the heavens; in the 1990s, during *tapsa* to Namsan, his students often joked that these pagodas should be seen as the highest in all Korea, insofar as their bases can be said to include several hundred meters of mountain. Yet Yun went on to align these Namsan pagodas with a specific national cultural relation to nature, rendering them as exemplars and prototypes, in the process bringing them into collaboration with other art historical claims and objects: "Might not [these pagodas'] clever conceptualization through which the humanly created *[in'gong]* and the natural *[chayŏn]* are connected be taken as one of the major strong points of our arts? We could say that in between the natural mountain and the artificial pagoda we approach the half-natural, half-arti-

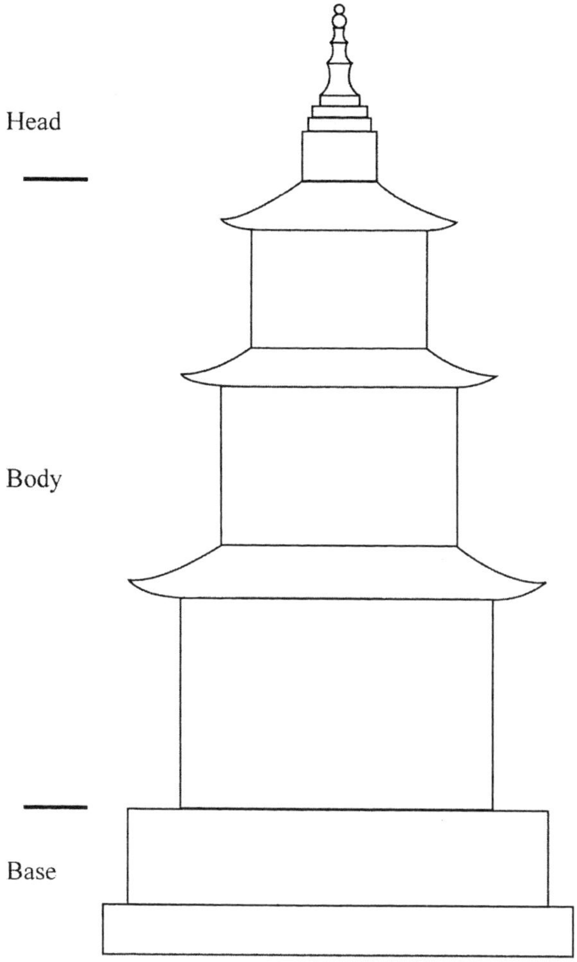

FIG. 8.1. Abstract form of a Korean stone pagoda *(t'ap)*, showing division of major sections. This is a three-story pagoda (common to Silla), because it has three body levels. Many of Namsan's stone pagodas lack the second (lower) base level; instead, the upper base level rests directly on the rock of the mountain. (Illustration by Jennifer Oppenheim, used with permission. This sort of diagram is common to the "curriculum" of Kyŏngju culture world organizations; see, for example, Yun Kyŏng-nyŏl 1993, 367, and 1994b, 423.)

ficial, and through the sublimation *[sŭnghwa]* of the mountain itself into a single pagoda we come to feel a more limitless space" (Yun Kyŏng-nyŏl 1979b, 28).

Yun thus used a classicism of form, an abstraction extant in art-historical scholarship, to define the concrete. He did not simply argue for the exceptional character of Namsan's pagodas, with their lack of a lower base level, but rather he staged a different sort of interobjective dialogue in which the generalized form framed their at once particular and nationally epitomizing relation of nature and artifice. A made scholarly knowledge object mediated cultural chthonicism.

Only momentarily, however, did Yun locate this consubstantiation of natural and humanly made, *chayŏn* and *in'gong,* in a single class of object before extending it to Namsan itself. One must, he claimed, understand the diversity in the number and scale of Buddhist statues, pagodas, and other cultural objects to be found at various locations in terms of the "necessary relations" *(p'ilyŏnjŏgin kwan'gye)* between the surrounding environment *(chuwi hwan'gyŏng)* and Buddhist doctrine. These relations "are not simply expressions of one pagoda or Buddhist statue but something more, a goal that can exist only in realization of harmony *[chohwa]* with nature." As a result of the achievement of this harmony, "although there are many temples and Buddhist relics on Namsan, there cannot be found one example of damage to nature." In Silla times, the boulders were transformed into Buddhas, and a previously simple place came to represent the Pure Land. "Therefore," Yun concluded, "in the case of Namsan the whole of the mountain *[san chŏnch'e]* must be seen as a relic *[yujŏk]*" (Yun Kyŏng-nyŏl 1979b, 29).

Many subsequent texts have echoed this formulation. Namsan itself is a relic or cultural object *(munhwajae),* and its cultural objectness inheres in its religious significance and in the relation of harmony between the made and the natural that it everywhere retains. Here lay the antithesis, arrived at via the very categories of historic preservation, to the dissolution of Namsan into simply a container for relics and cultural objects within the 1971 Kyŏngju Tourism Comprehensive Development Plan. Yun's project represents a local trajectory wherein the relationship with nature becomes an object of preservationist attention and an aspect of national character and being.

Going Home to Buddha Village

By the end of the 1970s, Namsan had become an object of localist study; by the end of the 1980s, it became an object of a much broader religious and social project. In no small part, this transformation can be attributed to the activities of Puch'ŏnim Maŭl, "Buddha Village,"[9] a largely lay religious group that arose in Kyŏngju but eventually found adherents and started branch organizations in other areas of the country.[10] Its leaders expanded upon Yun Kyŏng-nyŏl's appropriation of Namsan, connecting it with a broad vision of religious, environmental, historical, and social harmony.

One founder, Kim Tŏk-su, traces the origins of Buddha Village to a moment on a 1970 visit to Namsan when, in a flash of inspiration, he resolved to establish an eight-sided hall for veneration of the Buddha on the site. During the subsequent decade, he was involved in Buddhist lay groups, student groups, and propagation societies; in one letter he sent to a particular student organization in 1979, he used the phrase *"puch'ŏnim maŭl"* for the first time (Kim Tŏk-su 1984, 1). It was in another 1983 letter, however, sent from a dormitory in Saudi Arabia, where he was working on one of the large-scale Korean-directed engineering projects of that period, that the most oft-cited charter text of the organization appeared. Kim's verses of longing concluded,

> The thing that I want to have is Buddha heart *[puch'ŏnim maŭm]*
> The thing that I want to achieve is Buddha Village [Puch'ŏnim Maŭl]
> The one phrase that I want to shout is "let us revive Buddhism" (Kim Tŏk-su 1983)

Buddha Village was officially established in September 1984, shortly after his return. The organization styled itself as a "Buddhist movement [Pulgyo undong]"—a grassroots undertaking initiated by lay believers (although some individual monks did join) that sought revitalization from the inside out, a quality epitomized in humble precepts to offer devotion three times a day, give thanks before meals, and keep the poor in mind. Yet Buddha Village also called for reform, opposing "special privileges" *(igwŏn)* and "struggles for position" among monks and aiming for a return to the word and action of Buddha. The stated ultimate goal was the "Buddhification" of life and the transformation of the whole of the country into "Buddha Land" (Pulgukt'o) (Kim Tŏk-su 1984, 1–2).

A more in-depth doctrinal explication of the significance of the name of the organization was provided in the second issue of the *Buddha Village* newsletter by Sŏk Pŏb-u, himself a monk. Buddhism, he began, has throughout its history been concerned not only with the otherworldly but also with the "purification" *(chŏnghwa)* of the world in which humans live, an aspect evident in the sutras in the recurrence of such concepts as Pulgukt'o (Buddha Land or Nation), which might also be expressed as Puch'ŏnim Nara and Chŏngt'o (Pure Land). While Puch'ŏnim Maŭl might simply be seen as a variant of Puch'ŏnim Nara, Sŏk praised Kim's formulation for its "innocence" *(sobakham)* and its emphasis on the scale of real existence.

In the sweeping discussion that followed, Sŏk explored what and where

Puch'ŏnim Maŭl is, how it might be known, and how it might be achieved. He compared it to the notion of utopia in human and Korean history, as well as to specific concepts drawn from Buddhism (the Western Paradise of Amitabha), Christianity (the Garden of Eden and the Kingdom of Heaven), and even "communism" (the idea of a classless society). Yet he also emphasized that Puch'ŏnim Maŭl should not be seen as located only in a future place, but rather it must be conceived as an ideal or conceptual *(inyŏmjŏk)* space attainable within the existing life world (Sŏk Pŏb-u 1984).

Despite this distinction, as Buddha Village grew in membership,[11] Namsan became central to its social practice and a concretization of its idealized utopian space. As early as 1985, the group newsletter informed members that on the third Sunday of every month, "people who like Namsan" were meeting and conducting a "pilgrimage" *(sullye)* to the mountain. In 1992, this meeting spawned a formal branch organization to Buddha Village, the Namsan Sarang Moim (Meeting of Lovers of Namsan), which continued the Sunday pilgrimage while devoting the first two hours of each meeting to public information or environmental protection activities. Other more occasional events that attempted to involve a wider range of participants, such as the annual Namsan Sarang ŭi Nal (Namsan Love Day) that was organized from 1988, took on a similar format, combining *tapsa* or pilgrimage with efforts to collect trash left on the mountain. In 1986, Buddha Village held the first of several exhibitions of Namsan photographs; since then, Namsan postcards have also been available. Finally, alongside commentaries on sutras and excerpts from the poetry of Ko Ŭn, descriptions of Namsan's antiquities and personal accounts of journeys to the mountain continued to appear in the *Buddha Village* newsletter (e.g., Kang In-jun 1990; Kim Tŏk-su 1986).

What have the members of Buddha Village found on Namsan? Certainly one answer is the image of their own doctrinal underpinnings. In the *Buddha Village* newsletter *(Puch'ŏnim Maŭl)* during the middle to late 1980s, references to Namsan as "sacred ground" *(sŏngji)*, "the country of Buddha" (Puch'ŏnim Nara), or "a renowned mountain, held sacred as the Pure Land throughout the thousand years of Silla" are present. Other writings from this period would more explicitly access Yun's aesthetic of harmony epitomized in the close relation between natural and human creation. This text is from an unsigned book review that appeared in the *Buddha Village* newsletter (*Puch'ŏnim Maŭl* 1987, 11): "In the Buddhas of Namsan one can feel a soft and warm sense of intimacy that arises from the wondrous harmony between Buddha and nature. There is not the slightest feeling of distinction *[ijilgam]* between mountain and boulder and Buddha."

Harmony could also have a social referent, as became clear before a particular seated Buddha carving during a 1997 day spent with a senior member of Buddha Village and some other guests. The body of this figure is rendered in bas-relief, while the head is much more rounded and three-dimensional; the overall effect of a figure gradually emerging from the living rock anchored a narrative by our leader (Fig. 8.2). Before Buddhism came to Silla (in the sixth century CE), he explained, Namsan had been a center of animistic "rock belief" *(pawi sinang)*. After Buddhism was accepted, the royalty and aristocracy built grand temples such as Hwangnyongsa in the middle of the city. The common people, however, continued to come to Namsan. In their understanding of the religious transformation that had affected the kingdom, he continued, it was not that there had been boulders and now there were Buddhas. The thought was rather that the Buddhas had been there all along; they just had to be made visible. Yet another Buddha Village associate would deny to me later that Namsan had ever had a sole character as a site of popular religion. In recent decades, he explained, some outsiders had begun to write about Namsan as *minjung sŏngji*—the sacred ground of the *minjung*—but he rejected what he took as a politicized terminology and an exclusive understanding: Namsan in his description had never been a site of Minjung Pulgyo as *opposed* to Kwijok Pulgyo—of "people's" versus "aristocratic Buddhism." Instead it was a place where both elite and popular Buddhism were in harmony. A more senior member of Buddha Village put it in a similar fashion: "Namsan was a place where, day or night, the king or common people could come."

FIG. 8.2. The Buddha emerges from a boulder. Author's photo of the Large Seated Tatagatha Buddha in Relief of Lesser Naeng Valley (cf. Yun Kyŏng-nyŏl 1993, 268).

Mapping the Historiographic Sublime

Namsan is, ideally, an aesthetic experience for those who traverse its paths—one that partakes of all the senses but that from Yun's "Look!" on has usually been represented in visual terms. Kyŏngju culture world organizations have held exhibitions of Namsan photography; one winning entry, of a cliff-edge carved bodhisattva illuminated by golden morning sun against a backdrop of voluminous mist, was reminiscent of the masterpieces of European Romantic landscape painting to which the label of the sublime has been so readily applied. This resemblance is neither accidental nor superficial: The dynamic within the sublime, its circulatory doubling of absence and presence (Callon and Law 2004), is precisely what makes Namsan work in and as historiographic terrain.

For Edmund Burke ([1757] 1958) and Immanuel Kant ([1790]1952, 90–133), the experience of the sublime was one of awe, astonishment, wonder, tension—even initial terror or distress—in the face of unmasterable magnitude or might. Namsan's cultural objects are typically human in scale, but constituted as *itself* a relic, Namsan has been characterized in terms of its inexhaustibility. The unmasterability, immeasurability, and infinitude of Namsan are not of extension but of involution or depth, productive of wonder rather than awe. "Every time I go up Namsan, it is a new enlightenment *[kkaedarŭm]*," one Buddha Village member has written (Pak Chong-in 1996). The imbedded metaphor of the English word is, indeed, convenient here, for the Namsan sublime inheres in, and the mountain anchors stories of, the vagaries and revelations of shifting light—the winning photograph is only one example. Vision, enlightenment, and the materiality of light must in short be considered together: It matters for the politics of culture and relations of agency that undergrowth or rubble may conceal the bases of pagodas or the heads of chthonic Buddhas unknown, that in walking Namsan one passes, unavoidably and on the best of days, from the shadow of overhanging trees to the brightness of clearings.

Productively, Kant's sublime has a dynamic character. Unlike beauty, the sublime sets the mind in motion, and its pleasure derives ultimately from self-disclosure of the "mind of the judging Subject" ([1790] 1952, 104, 107). If we posit such a (male) visual subject confronted with infinitude, "a feeling comes home to him of the inadequacy of his imagination for presenting the idea of a whole within which that imagination attains its maximum, and, in its fruitless efforts to extend this limit, recoils upon itself, but in so doing succumbs to an emotional delight" (100).

Facing that before which the imagination fails, the viewer recovers the whole or totality only via reason, which throws precisely the "self-sufficiency" and "preeminence" of the faculty of reason into pleasurable relief, "mak[ing] us alive to the supersensible side of our being" (Kant [1790]

1952, 106–107). To Kant, the possibility of this pleasure in the sublime represented another proof of the synthetic a priori grounded in human nature. From Herder and Boas to Durkheim, many have sought in different ways to displace Kantian universals onto society or traditions or cultures. In the case of Namsan, the compulsion to and priority of the whole so central to the Kantian dynamic has an address neither in the mind nor in a super-additive generalized social (cf. Durkheim [1915] 1965) but in the determinate relations of spokespersonship in the insistence on the part of Kyŏngju culture world actors that Namsan itself is a relic or cultural object.

That the individual monuments on Namsan have value for Korean culture and Silla history is unchallenged. To the extent that Namsan is *additionally* accepted or prescribed as a whole—or as sublime in its infinitude—it can be taken as a "trap" (Gell 1996, 1998; Miller 2000), a site of leverage in which Kyŏngju historiographical agency is distributed. Namsan discloses itself at once aesthetically and historiographically—but not to all equally. On my first trip up its slopes with a Buddha Village member, he pointed out to me and the others along for the day the outline of a Buddha only barely visible on a worn boulder. It had been discovered only recently by another local historian, Song Chae-jung, who often walks Namsan's paths. As the story was recounted, it had been about four in the afternoon when, at *that* moment, in *that* low-angled light, something caught Song's eye that never had before. In the Kantian sublime, the partial encounter with the ungraspable threw one back upon the prior wholeness of reason. And indeed, each new revelation on Namsan, each vision and each new discovery, demonstrates both the partiality of any anterior estimation and the holism of the mountain itself: Look, Namsan has more to offer. But a prior whole may also structure partiality to political effect. Invested with value as inexhaustible, unmasterable, unfathomable in its true depth, Namsan is of a nature that rewards experience, repetition, and thus proximity *relatively* more than other sites. Namsan's holistic inflection mediates the translation of locality as dwelling or intimacy (Heidegger 1993; Raffles 2002b) and the possibility of familiarity borne of spatial proximity and frequency of traversal into a knowledge possessed by Kyŏngju historiographical actors that counts more widely.

Thus the infinitude of Namsan, and the sense of the radical impossibility of knowing it finally, have not barred but rather have underwritten Kyŏngju culture world efforts to render the mountain better for audiences in and beyond the city. Song Chae-jung himself, somewhat in the tradition of Yun Kyŏng-nyŏl and Buddha Village, has authored a new and more accurate map of its paths and cultural sites that has been widely reproduced and distributed. Song's interest in Namsan began in another luminous instant: In 1988, while driving toward Kyŏngju on the highway in the course of moving there to teach middle school, he was struck by the "golden light of the setting sun" shining off the mountain, so close to the city. Song's Namsan is mysterious and profound; it is a "spiritual mountain" that holds "the breath of our ancestors." It is an active force: "When I became obsessed *[mich'ini]* with Namsan," he has written, "it embraced me" (Song Chae-jung 1996).

Song had always liked geography and maps, nature and mountains. He suggested that the new map of Namsan, which took him five years, arose organically from his own practices of enjoyment. His habit has been to climb mountains alone, allowing himself to take turns and diversions, recording his thoughts and drawing his paths in a travel journal, out of which the pieces of the eventual synoptic map sprang. Song explained to me on another occasion that previous maps, many dating from the colonial era, contained errors or simply were not fine enough to be useful—their compilers had often not bothered to traverse the mountain themselves. His sense of the import of his own project dovetails with more widely circulated depictions of the uncharted quality of Namsan—in one of his own popular *tapsa* accounts, for example, prominent art historian Yu Hong-june (1999, 124–126; cf. 1993, 147–148) recalled searching futilely for hours for a particular stone Buddha in Namsan's unsigned interior. Yet simultaneously, Song implicitly denied that even his own carefully constructed map could be definitive or final. "There is not a single person," he wrote, "who knows Namsan completely!" (Song Chae-jung 1996, 33)

Other Kyŏngju culture world actors have sought in recent years to represent Namsan, as spokespeople, toward wider regimes of historiographical recognition. One aspect of these activities has taken place as if in response to Yu Hong-june's frustration. One 1997 public event largely sponsored by Kyŏngju's lay Buddhist organizations concluded with a long and grueling summer hike up Namsan. Three times we halted in front of permanent, weatherproof signs cemented into the ground of the sort that, while relatively scarce in the interior of Namsan, mark and explain officially designated cultural objects throughout South Korea. Unlike most, however, these had been paid for and erected by the Kyŏngju organizations that cosponsored the events of the day and bore text identifying themselves as such. Before each, leaders of the group performed a small ritual of unveiling—actually a reveiling-and-unveiling, with a cloth carried along for the purpose—and we offered bows while a sutra was recited. Like maps, such signs are also "signs" of engagement with public history and official heritage, on terms long negotiated and remade in and through Namsan as acceptable.

Conclusion

The Namsan of Kyŏngju local history developed in a gap left by academic historiography and the 1970s state transformation of the city. Thematically, it preferred multifarious "harmony" and humble Buddhas to the kings and generals of Yusin culture, but it insisted also that the mountain be taken as a whole, as a cosmos never fully graspable. The relation to the oppositional *minjung* discourse on culture that peaked in the 1980s was more complex. Many Kyŏngju celebrants of Namsan, themselves seeing it as a space attuned to the religious and cultural activity of common people, recognized the easy translatability of their concerns into *minjung* terminology. More often than not, however, they have forgone the translation, fearing the denaturing of their harmonious space, perhaps cognizant also of the precariousness of the achievement of local culture.

Many authors on locality and related themes have drawn spatio-epistemological typological contrasts between context-dependent, practical, supplementary *cum* resistant "local knowledge" and abstract(ing) imperial knowledge spatialities imposed from without (de Certeau 1984; Lefebvre 1991; Scott 1998). I am suspicious here of the implied romanticism of the local subject, but my substantive point is that the relation between these modalities must itself be understood as variable in character and effect, as dimension*ed* and not simply dimensional. The construction of Namsan as a sublime whole—its assemblage out of discourse, techniques of history, and embodied glimpses—mediates a specific relationship between intimacy and authoritative knowledge. One now arrives at Namsan along many paths, including those of official history and heritage and other basically centripetal projects, and Kyŏngju culture world actors now participate in making the mountain more available to outsiders. Yet Namsan is also a hole in the road, a point at which cultural depth becomes asymptotic to infinity. The agency of Namsan and its Kyŏngju sponsors together lies in their circulation among and between these disparate locations.

Notes

An extended version of this essay will be published in *Kyŏngju Things: Assembling Place,* forthcoming from the University of Michigan Press.

1. Yun passed away in 1999.

2. Originally, the membership tended toward doctors, lawyers, and middle or high school teachers. Over the years, with increased movement of nonprofessionals into the middle class, the proportion of members engaged in commercial and other sorts of activities has increased, but teachers are still the most prominent segment. Some museum professionals working at the Kyŏngju National Museum and (after these institutions came into existence locally) some university professors have also been members. The participation of women has also grown, but they have rarely held positions of authority within the group or written articles for the Friends' journal. One exception, however, can be drawn from the 1980s, when, as photographs make apparent, a Catholic nun held a position of importance.

3. It was "nationalized" and moved to a new building in 1975.

4. Most bibliographic sources give the name as Oba, as I have above, rather than Koba, as Yun Kyŏng-nyŏl does. Japanese names can often be read in multiple ways, and verbal forms can be difficult to reconstruct from written sources. In citing Yun's discussion, I follow his usage.

5. A rural administrative unit, larger than a village but smaller than a *kun* or county.

6. The word here is *"maŭl,"* which can be translated as "village," but it suggests something smaller, more informal, and more intimate than the legal village designation, *ri,* contained in "village head" *(ijang)* above.

7. A milky, fermented rice beer.

8. Harmony is, of course, a "recognizably Confucian concept"; Roger Janelli (1993, 108–109) describes its deployment in a very different South Korean sociopolitical setting.

9. This is correctly "Buddha Village" and not "Buddha's Village"—the leaders of the organization insist on the equivalent distinction, in Korean, between Puch'ŏnim Maŭl and Puch'ŏnim ŭi Maŭl.

10. Chung Byung-jo (1997) provides a general treatment of the contemporary Buddhist lay movement in Korea, while Buswell (1997) discusses lay organizations founded within monastic circles.

11. There are at present more than two thousand members.

Part 4

Geographies of the Margin

9

The Seoul Train Station Square and Homeless Shelters

Thoughts on Geographical History Regarding Welfare Citizenship

Jesook Song

This chapter attempts to make sense of homelessness and homeless policy in South Korea following the Asian debt crisis by examining the transformation of two symbolically charged physical spaces in Seoul: the Seoul Train Station Square (henceforth the Square or the Seoul Square) and the former Pangnim Factory (henceforth the Factory). I treat the Square and the Factory as spatial loci for evolving homeless policy and as sites for liberal and neoliberal historicity. Through the observation of changes in the spatial construction of these spaces, I trace neoliberal welfare ideology in relation to policies regarding homelessness and unemployment during the Asian debt crisis, or the so-called IMF crisis.[1] The label *ai-em-epŭ* (IMF) itself was a conspicuous marker of the crisis and neoliberalization. I specifically examine the social construction of the welfare citizen throughout the economic and political regime change sparked by the IMF crisis. Of historical importance on the crisis is that it coincided with the beginning of the Kim Dae Jung presidency. Kim's regime promulgated new social policies under the name of "Productive Welfarism" as part of an attempt to restructure finance, business, and government in accordance with recommendations from the IMF and the World Bank.

Chronicle of the Square

The Seoul Train Station is one of the largest and oldest transit centers in Seoul. It was built by the Japanese colonial regime as a hub for the many railroads connecting the Korean Peninsula to Manchuria, the route to the continent. In spite of many construction repairs, the original station buildings retain the typical architecture of the colonial period with a Renaissance style—a style that is otherwise rare due to the nationalist movement to eradicate colonial shadows.[2] Even after bus and subway became more popular modes of transportation in South Korea, the Seoul Train Station has remained one of the busiest spots in the city due to the countless bus lines and two major subway lines that stop in the station area. The Seoul Square has been a place for many rail passengers to rest or meet other people.

Typical of the Square at one time were scenes of people in transit, frequenting the small businesses targeted to them. Mobile bars and restaurants *(p'ojangmach'a)* were set up near bus stops and waiting areas to sell cheap food and alcoholic beverages, such as noodles *(kuksu* and/or *udong),* fried vegetables and seafood *(temppura),* seaweed rolls *(kimpap),* and Korean sake *(soju).* Mini food stalls were situated nearest the bus stops, selling bus tokens, gum, candy, and nonalcoholic beverages. Fruit and rice cake vendors set up their pushcarts and baskets in the entrance and stairs from the Square to the subway stations or walked among the people in the Square selling their goods. Until the early 1990s, the Square was also the site of mass demonstrations urging political action against authoritarian regimes.

Between 1998 and 2001, the Square changed.[3] During this time, it became well known as a place where enormous numbers of homeless people resided.[4] I visited Seoul in 1998, when the most common scenes were of numerous homeless people lying on benches and the ground in and around the Square or in the Station's underground tunnels. Some used newspapers as protection against the cold, and many were surrounded by bottles of Korean sake. While the Square had not presented the image of cleanliness before homeless people began to gather (sewage from mobile bars and restaurants as well as "night soil" of drunken men was common), during the IMF crisis city managers considered the Square to be even filthier due to the growing numbers of homeless people.

In the early morning and noon there were usually long lines for free meals, which were served by religious groups in several locations near the Square. In addition to the religious groups, a quasi-governmental welfare agency and at least two civil agencies without governmental aid provided support to the homeless at the Square. The quasi-governmental welfare agency was originally a civil group associated with the Anglican Church. The agency was then endowed and supported by the Seoul City government and became known first as the Homeless Assistance Center (Nosukcha Chiwonsent'ŏ) and later as the Homeless Rehabilitation Center (Nosukcha Tasisŏgi Chiwonsent'ŏ).[5] The Homeless Assistance Center provided medical examinations to homeless people who registered for a Homeless Identification Card (Nosukcha K'ad).[6]

One civil agency was the Humanitarian Practice Medical Doctors' Association (Indojuŭi Silch'ŏn ŭisa Hyŏpŭihwae), which assessed the health condition of homeless people coming to the Square (Park Yong-hyŏn 1999). The other was a private welfare agency run by a renowned South Korean comedian. This agency operated a bus called the Telephone of Love at the Square. Telephone of Love workers served homeless people who were willing to complete a questionnaire that asked about their family situation, hometown, employment history, and length of time living on the street. In return for this information, the agency provided clothing, toiletries, and nonperishable food.

During the day, many homeless people left the Square as it became busy with passengers, or they were driven out by station guards. By November 1998, homeless people were not supposed to be in the Square even at night, as the Seoul City government made it illegal for street people to be in public places. Instead, they were directed to go to a homeless shelter. However, there were still noticeable numbers of homeless people staying in the Square. They received charity goods and free meals from various humanitarian groups, such as the Telephone of Love, but they were reluctant to follow city policy to reside in homeless shelters because the condition of entering those shelters required disclosure of personal information and restriction of certain freedoms, such as drinking and smoking.

Through the Square, we can see two contradictory sides of South Korean society at the time: the disenfranchised citizens—unlucky to lose their jobs and homes—and the services established to support them; and the commuters who still possessed the means to travel *through* the Square, stopping only momentarily at businesses catering to their needs. The fact that homeless people were driven out of public places during the day and were later entirely prohibited to stay reveals how homeless policy was actually designed to protect "regular" citizens who might be offended or harmed in some way by the presence of homeless people. Narratives of government managers during this time suggested that homeless people might threaten the safety of "regular" citizens or become violent through riots. Policy was developed so that the homeless would be removed from the eyesight of the "regular" people who passed through public places, thus spatially segregating Seoul's citizens.

My visit to the Square in 1998 revealed another aspect of the spatial construction of the welfare subject not only as different from the regular citizen but as a gendered construct. There was a police station at the outskirts of the southeast corner of the Square. Across from the police station was a small, old, one-story building with a sign that read "Women's Welfare-Counseling Center" (Yŏsŏng Pokchi Sangdamso). I wanted to know what the center was about, but its doors were firmly closed. As I looked around, I saw a policeman smoking outside the police station. He was looking in the direction of the Square where many homeless people could be seen. From everything I had heard, policemen aggressively drove homeless people from public places or urged them to go to homeless shelters. But this officer did not seem to negatively stare at the homeless people. Rather, his eyes showed a meditative mood, like an old man smoking at the entrance to a village with little to do but think and silently observe neighbors' trivial actions.

When I asked him if the Women's Welfare-Counseling Center (WWCC) was temporarily closed for the day, he told me that the center had been closed for months. He explained that a woman (the only staff person at the center) used to bring food to the police station but that without any notice, she had stopped coming. He seemed to have enjoyed her visits and was perplexed by their abrupt ending. Shortly after she stopped appearing at all, the center was closed. He did not know why.

Through my conversation with him, I learned that the WWCC had been administered by the district office where the train station was located. In an effort to find out what had happened to the center, I called Seoul District Offices and found out that they had laid off the only person responsible for women's and family welfare, a position created in 1997 under the Kim Young Sam regime.[7] The person discharged was the woman who worked in the WWCC. One employee alluded that she was one of the casualties of "voluntary layoffs" *(myŏngyet'oejik)*.

Apparently the local government had been under pressure to "restructure" the administrative system immediately after the South Korean government received the IMF bailout funds, when South Korean bureaucratic machinery in addition to banks and big corporations became primary objects in need of downsizing in the name of efficiency and flexibility for global market competition. Thus, the Seoul government laid off workers who did not have good records or whose duties were not considered necessary. Indeed, there was a tendency to lay off female workers first, including government employees—an illegal discriminatory practice. The central government administrator for women's affairs in the Ministry of Health and Welfare did not know of this incident when it happened. When she heard of it later, she lamented, saying, "We told them [the local government and districts] not to discriminate against female employees, especially public social workers. But see, they didn't listen to us."

As I made my inquiries with district social workers, I also tried to gather data on the situation of deprived women during the crisis. I asked how many times they had received phone calls or visits from women whose destitute situations or homelessness could be a result of the IMF crisis and if they knew of any homeless women's shelter that the city financially supported. I could not learn anything from the workers I spoke to, as the laid-off employee was the only person responsible for maintaining this information. Not only had the service for women been spatially relegated to the outskirts of the Square, but the one worker responsible for this service had been eliminated—and with her went all records of the female welfare subjects.

When I revisited the Square in October 2001, the former Women's Welfare-Counseling Center had been transformed into the Homeless Counseling Center (Nosukcha Sangdamso), now run by a religious organization. The changed title and function of the district office building from "women and family" to "homeless" service is significant, as it signals a welfare subject constructed differently in successive regimes—women/mother in the Kim Young Sam regime and homeless men in the Kim Dae Jung regime.

In general, men in need were considered more urgent and deserving welfare recipients than women in need during the crisis in relation to the social anxiety of the faltering middle class, as a result of mass layoffs of life-employed middle-aged men from big corporations (Kim and Finch 2002; Song Jesook 2006). Even those considered "deserving," however, were met with meager support (see Mr. Ku's account below). As before, the center was not staffed during the day, and the building, still shabby and locked, stood out in stark contrast against the rest of the Square, which was radically changed.

My visit in 2001 coincided with the time when it was frequently reported in mass media and government rhetoric that the crisis was over. The Square had been rearranged as a huge, clean parking lot for the customers of the Seoul Train Station. I could not find any mobile bars or restaurants *(p'ojangmach'a)* within the barricades of the parking lot or nearby bus stop area. There were only a few remaining fruit vendors with pushcarts or baskets, a change that seemed linked to the fact that South Korea and Japan were cohosting the World Cup Soccer tournament in 2002.

Before the IMF crisis, the government had attempted to get rid of mobile bars and street vendors in most of Seoul's public places with the intent of creating an image of Seoul as a clean metropolitan city—an image that would appeal to foreign travelers. During the crisis, however, there were several massive demonstrations and strong resistance from the owners of the mobile bars. They charged that the government was "suffocating" the poor working-class people, who were managing to run marginally profitable businesses in such a dire time. With increasing concern about social instability due to mass unemployment following the IMF bailout, the Seoul City Office withdrew its order to remove the mobile bars in an attempt to stave off further unemployment. Originally, veterans and poor families were granted limited permission to run their businesses, but these regulations changed with economic fluctuations. During the crisis, it was very common to see college students and unemployed people starting mobile restaurants in residential areas. In 2002, when the World Cup was held in Korea and the crisis was officially over, street vending was again restricted. Shifting policy on the street vendors is a good example of how marginal welfare citizens were selectively dealt with at different times during and after the crisis. Now, a large new gallery and souvenir shop displayed this upcoming event through its panoramic glass storefront. Situated prominently in the Square and boasting huge glass windows on three sides, the gallery shop focused on multimedia effects: It displayed various products with the 2002 World Cup logo and mascot and aired either commercial advertisements or specially produced films by the South Korean government on multiple TV monitors attached to the glass walls. It also blared the World Cup's theme song across the huge plaza. Against the dramatic backdrop of the newly transformed Square, the district building told an important story: Despite surface changes to policy, the practice and implementation of new welfarism by local government maintained the shabby exterior of the building and an unwelcoming image—*a firmly closed door.*

I now turn to a more extensive analysis of the implementation of new welfare policies—a process that further demarcated the new welfare subject both from the regular citizen and from the "othered" homeless.

The Conversion of Pangnim Factory: Homeless People as IMF Subjects

I have introduced the increased presence of homeless people in the Square as a novel phenomenon of the IMF crisis. I revealed how the use and transformations of the Square parallel how the welfare citizen was constructed as other from the "regular" citizen. I have also demonstrated the gendered spatiality of the Square and shown how successive regimes removed women from the image of the welfare subject. In this section, I examine more closely the construction of the "deserving" welfare citizen during the IMF crisis, which resulted in the distinction of "IMF homeless"—supposedly short-term street-living people—from *purangin* or long-term street-living people. I will demonstrate how the emergence of "IMF homelessness" in Seoul embodies a larger transformation of South Korean welfare regimes, while the evolving discourses and practices related to homelessness can be seen as a sociocultural marker of the materializing neoliberal welfare state and citizenship. In a strict sense, homeless policies were emergent at the outset of the crisis in the winter of 1997–1998, before the announcement by President Kim Dae Jung of a blueprint for "productive welfarism" in July 1998. However, the implementation of homeless policies through the Seoul City Office was advised and supervised by the president's office from the beginning of Kim's regime in early 1998. Kim was already responsible for the administration's response to the crisis after wining the election in 1997—at almost the same time as the crisis broke.

In particular, when the central state and Seoul City Office administrations first recognized homelessness as a major emergency welfare concern, only "IMF homeless" became legitimate welfare subjects—the measures of employability and potential for "rehabilitation" demarcating proper subjects from *purangin.* This selective legitimization of homeless people is symptomatic of the recently introduced neoliberal welfare ideology's selective reconfiguration of its "appropriate" citizenship. Unraveling the way in which the South Korean state administration legitimized selective welfare citizenship, I focus first on the role of the media and higher-level government in the initial construction of IMF homelessness and then on the implementation of policies relating to homelessness by Seoul City. Government administration is just one of many social and political domains that effected the neoliberalization of South Korean social governance (or governmentality); however, it is not an exaggeration to assert that on the brink of the IMF crisis, the Kim Dae Jung administration's implementation of "productive welfarism" was a crucial impetus for amplifying the neoliberalization of South Korea's state policies. By introducing the way in which quasi-governmental agents were actively mobilized in the implementation of homelessness policy through the city's shelter system, I will elucidate an aspect of neoliberal governmentality whereby the state acts upon nonstate agencies. I examine how the dichotomization of Seoul's homeless people was spatially executed and reinforced through the city's shelter system, highlighting the transformation of the former Pangnim Factory.

Mass Media and Politicians: IMF Homelessness

During the crisis, two terms evolved to distinguish "IMF homeless" from other homeless people: *"ai-em-epŭ shiljik nosukcha,"* which translated means "IMF laid-off homeless"; and *"ai-em-epŭ hyŏng nosukcha,"* which means "IMF–style homeless." The term *"purangin,"* a derogotary term for street people, was used for non–IMF homeless, while *"nosukcha"* (homeless people) was used to designate people in the street without any negative connotations (Song Ch'ang-sŏk 1999).

The dichotomized categories of IMF homeless versus *purangin* were widely circulated by mass media and governmental managers. Seeking publicity, politicians and upper-level officials paid visits to homeless people, accompanied by photographers. Despite the fact that homeless people resided in other public spaces, these visits targeted those staying in the Seoul Square, who soon became the tragic "face" of the IMF crisis.[8] Beginning in the winter of 1997, when the crisis broke, mass media (including newspapers, special report television programs, dramas, and novels) often dealt with the subject of IMF homelessness, frequently showing scenes of the Square.[9] The homeless were portrayed as casualties of the IMF crisis, resulting in donations and funds for the IMF crisis that targeted those people on the edge of unemployment and homelessness.[10] One reporter's comments summarize the image that was constructed of IMF homeless versus *purangin:*

> Among unemployed homeless people, there are many who used to sleep in the temporary kitchen of a construction site or in restaurant halls downtown. It took at most one or two months for them to end up on the street after becoming unemployed. There are cases of people becoming homeless immediately after becoming unemployed. However, we should not confuse these people [unemployed homeless people] with *purangin,* who wander because they *cannot adjust to a work place and family life.* Unemployed homeless are people who would be able to settle down in minutes, as long as they have work with which to feed their families and have a place for the family to spend the night together (*Han'gyŏreh* newspaper, September 18, 1998, my translation and italics).

Social workers who met and lived with homeless people contested this clear division and challenged the image of IMF homeless as necessarily easily "rehabilitated" and having a middle-class background. Even the police officer I spoke to in 1998 challenged the distinction: He maintained that before the IMF crisis, there had been a lot of homeless people in the Square.[11] Although there were more homeless people living in the Square during the crisis, he thought most of them were the same kind of wanderers who lived a street life before the crisis—coming and going as they pleased. Showing his suspicion about sharp differences between recent homeless people and precrisis homeless people, his narrative challenged the newly developed view of homeless people as IMF homeless.

Nevertheless, the government officially insisted that homelessness coincided with the massive unemployment following the IMF bailout and built its homeless policy upon this false premise. The government publicized that it would provide support via work programs and shelters for the IMF homeless, who would soon be rehabilitated. On the surface, government officials denied that homelessness had really existed prior to the IMF crisis; however, pressure from workers in the field forced them to address the fact that there were homeless people who did not fit the criteria for IMF homeless. Thus, the welfare and shelter systems developed with the external appearance of supporting only the IMF homeless, while at the same time removing others from public sight via the city's shelter system.

The Spatial Construction of Homelessness in the City's Shelter System

The homeless shelter system in Seoul, established shortly after the emergence of the crisis, consisted of two categories of shelters: Houses of Hope

(Hŭimang ŭi Chip) and the House of Freedom (Chayu ŭi Chip). Houses of Hope were small homeless shelters where only "IMF homeless" were supposed to stay. The benefits of staying in Houses of Hope included being paid, being able to work at Public Works Programs, and getting free meals and a place to sleep. Because of the government's policy to support only IMF homeless, field workers were required to "sort" IMF homeless from others. Thus, the House of Freedom became the first stop for all those entering the shelter system. Here they would be assessed, and those who qualified as IMF homeless would be admitted to the Houses of Hope.

The House of Freedom is a huge building located in the Kuro industrial area of Seoul. The building, now owned by Seoul City, used to be called Pangnim Factory. It was one of the first and biggest textile factories in the developmental period under the Park Chung-hee regime and once symbolized South Korea's economic success, built on light industry in the 1970s. The 1970s in South Korea were a time when the labor movement began its resistance against the exploitation of primarily female laborers by factory owners who supported the Park regime.[12] The Pangnim Factory is the site of one of the most memorable labor struggles of the time, when the appeal of female laborers for humane working conditions (i.e., an end to unpaid night labor) won wide support among labor and social activists (through Yŏngdŭngpo Sanŏp Sŏnkyohoe or the Yŏngdŭngpo Industrial Mission). The triumph came through consecutive demonstrations and violent police actions against demonstrators, yet despite their struggles, female laborers were all fired through an alliance between the Factory owner and the Park regime.

By the 1980s, heavy industry, such as automobile manufacturing, began to represent South Korea's national prosperity throughout the world, and light industry had declined to the extent that the Factory shut down (Cumings 1997; Kim Eun Mee 1997). The building stood empty for many years before being reopened in 1998 as host to thousands of homeless people.[13] The site once symbolizing South Korean enterprise, sucking laborers' blood and kicking them out of the workplace, was now providing shelter to multitudes of supposed IMF casualties with the aim of sending them to work. The striking and ironic contrasts observed in the history of Pangnim Factory include its identity as a locus for economic prosperity in the 1970s and its current status as a shelter for the most destitute members of South Korea's citizenry; and its identity as a place notorious for the exploitation of laborers in the 1970s, now known as a "shelter" for jobless and homeless people.

I visited the House of Freedom in 1999. I had arranged an interview with Mr. Ku, a quasi-governmental employee who ran the shelter and the Homeless Rehabilitation Center (HRC)—a mid-level civil agency mandated by the government to assign homeless people to the Houses of Hope. In the management office where we waited, we sat with a few relatives of homeless residents. Many college students studying social work went in and out of the office, making notes on the board or looking for visiting family members. There were ceaseless announcements over a loudspeaker requesting that social workers look for homeless people in order to counsel them or to arrange meetings with visiting family. Some social workers from the Houses of Hope broadcast announcements to pick up those who passed as IMF homeless.

The House of Freedom's former life as a giant textile factory made it spacious, but its industrial past and its long vacancy had left the walls and floor ugly and bare, giving it an empty and chilly feeling. The constant announcements sounding through the air of the old factory building reminded me of an old train station lobby in which people come and go quickly or sojourn shortly, much like the Seoul Train Station waiting room. The high ceilings and empty space caused the announcements to echo deafeningly through the air. These echoing sounds might have seemed familiar to some street people, who were used to staying in various train stations. The House of Freedom thus recalled the Seoul Train Station, but with the added dimension of being legitimized as a space in which homeless people were institutionalized and sorted into "governable" or "ungovernable" welfare subjects.

Mr. Ku arrived to our meeting late as he had been held up in an emergency meeting regarding violence among the residents of the House of Freedom. He introduced himself as a once-devoted student activist in the early 1980s and seemed more comfortable talking with me after discovering that I was also involved in the student movement during the 1980s, even though I was junior in generational terms of activists. In a context where the "welfare state" was introduced to implementing neoliberal policies in the late 1990s, it is interesting to observe how student activists from the 1980s became leaders of nongovernmental organizations involved in the government-initiated implementation of welfare society. Particularly interesting is the change in social reformists' rhetoric from one decrying dictatorship to one espousing civil society as a background of the amplification of neoliberal society in South Korea. When the subject turned to IMF homeless, Mr. Ku provided clear insights into the problematic construction of the IMF welfare subject. He attested that the government had made an error in ordering social workers to distinguish IMF homelessness from long-term homelessness, indicating that the distinction was not so clear, and that the rehabilitation of any homeless people would not soon be accomplished as the government had promised:

> The city [managers] might think that they successfully arranged things regarding homeless issues since they perfectly managed homeless people

> so they would not be seen by the eyes of the citizenry. However, it is naïve for the City managers to think that IMF homeless will immediately return to society if only the City offers them a place to sleep and a temporary job for the winter. Problems of homelessness have only just begun. . . .
>
> In order to analyze fundamental problems of producing homeless and to categorize them in an appropriate way, it is necessary to counsel them in depth for a long time and to conduct qualitative research. But there are few specialists of homeless issues, and any research budget is the first to be cut by the governments, nowadays. This is probably because the city doesn't consider the homeless as regular welfare subjects (Mr. Ku, manager, House of Freedom, interview by author, Feburary 20, 1999).

According to Mr. Ku, this was a meaningful time for him and other graduate students of social work to be able to study homelessness as the newly emergent social welfare subject. However, in addition to his enthusiasm for the possible academic contribution, he also had pragmatic concerns about managing issues of homeless people. One big problem for him was the government's requirement that he identify and provide services only for "IMF homeless." He said that many homeless people had been on the street and in public places like the Seoul Train Station for a long time—more, in fact, than people who had recently begun to live on the street. So, the requirement to remove all homeless from the city's public places to the HRC was in tension with the city's intention to provide services only for "IMF homeless." Mr. Ku said it was difficult to distinguish IMF Homeless from *purangin,* and he indicated that the problems ran much deeper:

> The government might reduce the number of homeless people through short-term benefits for them—shelters and work opportunity at Public Works Programs (PWP)—but those benefits rendered an explosion of the population residing in these shelters, and now governments face the trouble of managing them. Homeless problems should be solved with a long-term plan. The degree of benefits for people who received governmental subsidies for livelihood is lower than that of homeless people. So, House of Freedom intentionally doesn't provide a benefit for PWP [different from Houses of Hope]. But without the benefit of PWPs, homeless people tend not to come to the homeless shelters (Mr. Ku, manager, House of Freedom, interview by author, February 20, 1999).

At the one-year celebration symposium of HRC, Mr. Ku distanced the HRC from governmental policies and representation of homeless issues (Sŏh 1999). He announced that policies for IMF homeless were limited in their ability to address the needs of the rest of homeless people and revealed shocking statistics. He maintained that only 20 percent of the homeless population could be classified as "IMF homeless," dealing a blow to the governmental premise that homelessness coincided with the economic crisis.

Mr. Ku insisted, "Homeless problems in our society have been latent because of a lack of a social safety net and a meager welfare system. Therefore, it is necessary to reexamine the direction of homeless policy in a long-term plan." This public announcement was a blow to the government, which had attempted to dismiss the historical and structural aspects of homelessness for which government could be responsible by underlining the temporary nature of homelessness as caused by the IMF crisis. HRC's presentation of statistics, traditionally the authoritative domain of bureaucracy, was especially devastating.

Governmental homeless policies created the "IMF homeless" as the only legitimate welfare subject deserving of public services in the era of the IMF crisis. *Purangin* were othered, as policies were predicated upon the assumption that homelessness emerged only due to the crisis. These assumptions were contested by field workers such as Mr. Ku and the policeman who had been in the Square for some years. While the Kim Dae Jung government's welfare state conveyed an image of successfully "guaranteeing the minimum standard of living" through its inauguration of homeless measures, the administration's "guarantee" did not apply to all homeless people; instead, it established the state as mediator by targeting certain homeless people whose employability was proven. The Kim administration aimed to mobilize self-governable welfare citizens who were not dependent on the state—a goal that I identify as a product of the emergence of neoliberal welfarism in South Korea. The pursuit of disciplining welfare citizenship toward workfare or postwelfare citizenship has not had much impact on the amelioration of disenfranchised people, such as long-term street-living homeless people, even when a homeless policy was systematically implemented for the first time in Korean history. Spatial constructions in the Square and the Pangnim Factory (the House of Freedom) embody sociopolitical construction of "deserving citizenship" (i.e., homeless vis-à-vis "regular" citizens; women in need vis-à-vis homeless men; IMF homeless vis-à-vis *purangin;* and textile factory with a history of exploiting laborers vis-à-vis homeless shelter).

Notes

I am grateful to Richard Roy Grinker for allowing me to use similar content from my article in *Anthropological Quarterly* in this edited volume. I also would like to thank Timothy Tangherlini, Sallie Yea, Laura Nelson, Lynn Clark, Hon-Yee Choi, and Ken Kawashima for showing interest in this essay and providing comments.

1. In referring to the Asian debt crisis, I will use the phrases "IMF crisis," or "the

crisis" interchangeably. South Korean people tend to refer to the crisis uniquely in relation to the IMF (International Monetary Fund: the supranational financial institution that "bailed out" South Korea's debt). For instance, *"ai-em-epŭ satae"* (IMF crisis) was the most popular label during the crisis—*"satae"* means emergency or warlike crisis in Korean. *"Ai-em-epŭ"* (Korean for IMF) alone was another popular label used to refer to the crisis in South Korea, despite the fact that many Koreans know the IMF is actually an institutional title. For instance, *"ai-em-epŭ ga watta"* means "the IMF crisis arrived," and *"ai-em-epŭ ttae"* means "when there was the IMF crisis." Finally, *"ai-em-epŭ sidae"* (IMF period) indicates a more retrospective, time-centered concept used after 2001, when the crisis was officially recognized as ended. As in the case of discourses surrounding "IMF homelessness" and "IMF unemployment," it denotes the involvement of a foreign power in the national crisis and its negative impact. On the one hand, it does not necessarily mean that South Koreans blamed the IMF as the sole factor of the national emergency. However, their way of naming the economic crisis might have been both influenced by and reinforced the popular thought that the national emergency was caused by a foreign power interfering with South Korea's sovereignty.

2. It was designed by a German architect in 1925. See Daos Korea Co. (2004).

3. Although the crisis began in November 1997, its impact was felt mostly from 1998 on.

4. In August 1999, the city estimated that there were 2,000 "IMF homeless" in the Square. In the winter, the number increased to 4,000.

5. The Homeless Rehabilitation Center (HRC) was established in 1998 as a compromise plan after Seoul City spent several months attempting to deal with homeless issues (Park Kŭn-ae 1998). The replacement of "Assistance" with "Rehabilitation" in the title is symbolic of the transformation of welfare regimes from Poor Relief to the neoliberal workfare regime.

6. The registration process of the governmental agency in the Square was difficult for homeless people to endure. Social workers assisting street people in temporary homeless shelters said that receiving and carrying a "Homeless Identification Card" was a very degrading experience for many homeless people. There were also incidents where police booked homeless people after using the identification card to inquire into their background, further damaging homeless people's trust in the shelter workers.

7. The position of women's welfare counselors was created at the end of Kim Young Sam's presidency (1993–1997), in concert with laws aimed at preventing domestic violence and protecting the victims. However, local police officers did not enforce the new laws even after they were implemented. They continued to exhibit conventional attitudes toward domestic violence, maintaining that public officers should not intervene in domestic violence because it is a "family" matter. As a result, most police reports of domestic violence ended up giving more authority to the view of male patriarchs.

8. For example, Ko Kŏn, the mayor of Seoul, stopped by on September 21,

1998 (Kwŏn Hyŏk-ch'ŏl 1998); Kim Po-im, the Health and Welfare minister, visited on June 2, 1998 (Ahn Ch'ang-hyun 1998); and Ch'a Hŭng-bong, the next minister, came by on July 21, 1999 (Sŏn Tae-in 1999). Also there were visitors from the ruling party and the opposition party. It is notable that homeless people in the Station complained when reporters took pictures of them with Ko Kŏn for political use.

9. See, for example, *Road Movie* (Kim In-sik et al. 2003); *Sad Temptation* (Pyo 1999); *Men Whom We Loved* (Yi Na-mi 1999).

10. Yi Wŏn-jae reports that there were even donations from the transnational Korean community in the United States (Yi Wŏn-jae 1998; see also Yi Chang-kon 1998).

11. Interestingly, he did not use any particular word to indicate homeless people, nor the recent categories "IMF homeless *(nosukcha),*" nor the longstanding "*purangin.*" Rather, he used general indicators for persons—"*chŏ saramdŭl*" (these people).

12. This is the context in which Chon Tae-il committed suicide by self-immolation after requesting an end to labor exploitation in Pyŏnghwa Market, commemorated as a cornerstone of the South Korean labor movement history.

13. The House of Freedom was originally prepared to accept 300 street people, but on the opening day 1,200 people crowded into the building (personal interview with Mr. Ku, manager of the House of Freedom, February 20, 1999).

10

Cyberspace and a Space for Gays in South Korea

MICHAEL J. PETTID

CULTURE IN CONTEMPORARY South Korea seems in a state of continual change and reassessment due to the influx of new or outside value systems and of a reevaluation of what are thought to be traditional values. As a part of this process, many individuals are seeking to either establish or redefine both personal and larger, group-orientated identities. Of course, formulation of identities and creating spaces for personal exploration is not a matter free from criticism, and further, it is an area that individuals are constantly reconfiguring and adjusting. One particularly important stage of contention has been understandings of sexuality, especially sexuality beyond the more-or-less socially expected roles of men and women.

We should begin this investigation by acknowledging that discussion of sexuality, and by extension the body, is in fact an inquiry into a "site of action" (Pile and Thrift 1995, 19). In that the body is a concrete and oftentimes contested space, there arises a need to comprehend how this space is demarcated and defined. Understanding the subject of the body requires acknowledgment that various regimes of power influence how we see ourselves and others and also shape the spaces where we are permitted to be actors and the means by which we can express ourselves. Limiting forms of power can concern knowledge, class, gender, and sexuality among other structures and contribute to the formulation of identity, both positive and negative. When such hegemonic power regimes are resisted, the body itself can become a site of insurgency and a place where boundaries are reshaped and the self transformed. Thus, a significant degree of agency can be realized in the control of one's "site of action," notwithstanding various outside restraints.

The chief focus of this chapter is how gays in South Korea contend with larger society as gays and, peripherally, how heterosexual identities are brought into question by the outward presentation of a "gay" space. The cardinal factor that precipitates hardship for gays—and all lesbian-gay-bisexual-transgendered (LGBT) persons—is a general intolerance and, perhaps, an overall ignorance of alternative lifestyles by official or public Korean society. Such difficulties arise from the challenge presented by these subjects who transgress dominant boundaries. The performance of nonhegemonic sexualities can be characterized by tension between "discourses of regulation and resistance," and these discourses are "articulated in different ways at different times and at different places" by the actor (Bell and Valentine 1995, 143).

For gays in contemporary South Korea, management of time-space boundaries concerning the outward performance of their sexual identities is crucial to minimizing the social hardships that can cause loss of economic, educational, and even creative opportunities. Outward rejection of traditional heterosexual roles is not always possible or wise in a society that largely rejects what are viewed as "nontraditional" sexualities. Compounding difficulties for gays are the still-strong Confucian ethics—most prominently filial piety and the importance placed on lineage—that shape many aspects of contemporary South Korean society. Finally, gays also must overcome the prejudices against such a lifestyle found in Western scientific and religious discourse.

Historical Considerations

It is helpful to examine the views on homosexual activities in past Korean societies and how these changed with the introduction of Western scientific and religious discourses.[1] While there is little doubt that there were same-sex sexual relations in premodern Korea, these were not labeled as "homosexual"; there simply was no such category to place people.[2] In the West, the discourse on homosexuality was not formulated until the late nineteenth century, when a transformation occurred.[3] Michel Foucault (1990, 43) put it this way: "The sodomite had been a temporary aberration; the homosexual was now a species." With this change, homosexuality became something to be "cured" and was viewed as a tremendous "sin" against nature (Foucault 1990, 101). As I will demonstrate, the discourses on homosexuality in Korea have followed a strikingly similar course.

While records concerning same-sex relations are fragmentary at best in premodern Korea, there is sufficient evidence to permit a conclusion that these relationships were accepted or at least understood as a part of life and, while not necessarily embraced as desirable or proper, did not require treatment or punishment.[4] For example, lesbian relations between palace women must have been extremely common, as the practice of "pairing up" as a couple even had a special euphemism *(taesik)* in the Chosŏn period

(1392–1910) (Kim Yong-suk 1987, 73). One characteristic of the institution of the *namsadang*—male troupes of itinerate entertainers—is the homosexual relations between younger members, who dressed as girls, and elder members; the *namsadang* is also known to have supplied male prostitutes to the areas it traveled through (Shim 1994, 5:408–411). Furthermore, homosexual practices seem to have been quite common at Buddhist temples, even until recent times. Gregory N. Evon (2001, 14–15) has written that while there was heated discussion about celibacy and Buddhist monks in the colonial period (1910–1945), this was only in regard to *heterosexual* sex—"homosexuality simply elicits no comment."

This is not to state, however, that homosexuality was encouraged or celebrated by official Chosŏn society; rather, at times it was used to criticize political opponents (Pratt et al. 1999, 167).[5] Those who were too inclined toward same-sex relations were seen as, at the very least, being out of touch with their prescribed social role. One such example is that of Grand Prince Chean (1466–1525), the son of King Yejong (reigned 1468–1469), who is recorded to have had distaste for women. While the prince faced no official censure, the following commentary reveals how many would have looked upon him: "In general, the desire between men and women is something that is present from birth and such human desire cannot be blocked. [Prince Chean] held throughout his whole life that women were dirty, and he always avoided them. If [such actions] are not idiotic, what is?" (Ŏ 1971, 489, 758).

Despite a more-or-less tolerant attitude by those around him—the prince never was censured for his behavior—his actions made him the subject of humor and contempt. The above description also demonstrates that such a condition was something understood as inherent in one's sexual nature and not something that could be altered or treated.

Perhaps contributing to the propensity to "look the other way" in the case of same-sex sexual relations in Chosŏn was the Confucian view of sexual matters as being a part of the private sphere. Episodes concerning sex were not a part of the public discourse, and Confucian ethics highly discouraged bringing matters concerning sexual practices into the open (Pettid 2001, 61–86). When such episodes were brought to public light, however, the consequences could be severe, as in the case of Lady Pong.

Lady Pong was selected as a wife of the crown prince in 1429, but soon after the marriage the relationship between her and her husband became strained. In a later petition to the throne, Lady Pong is accused of a wide scope of "crimes," including those deemed immoral such as loudly singing male love songs to the palace maidservants and spying upon women in the toilet. These matters aside, her greatest transgression was that committed upon the maidservant Sossang, recounted in the dynastic record: "This past winter solstice the Queen summoned me to her room; all the other maidservants were outside her room. She demanded that I lay with her, but I refused. She then forcibly tore about half my clothing off and pulled me behind the folding screen; there, she took off the rest of my clothes and forced me down. Then, just like a man having sex, she trifled with me" (*Sejong sillok* 75:7b–9b [1436/10/26]).

This incident, along with various others, was used as evidence for deposing Lady Pong (*Sejong sillok* 75:7b–9b [1436/10/26]; 75: 14a–15a [1436/11/07); 75: 26b–27b [1436/12/28]). Other accounts echo the above, and Lady Pong's blatant sexual behavior was deemed something that could not be tolerated in the royal palace. Her open pursuit of same-sex relations was the reason for her official censure. Thus, in contrast to the description concerning Grand Prince Chean, Lady Pong's public flaunting of an "abnormal" behavior—at least to Confucian eyes—was the source of her demise.

Further obscuring premodern discourse on homosexuality is that engagement in a homosexual or lesbian relationship was not an identity-defining matter. In premodern Korea there was simply not a homosexual identity or label with which to "mark" an individual as different or abnormal. Sexual activities in general were considered matters of a private space. In Chosŏn Korea, there were no words, such as *tongsŏng aeja,* to label those with alternative sexual identities. This would have to wait until the introduction of Western science and religion in the late nineteenth century.

The late nineteenth century in Chosŏn was a tumultuous time. The political foundations of the dynasty had been thoroughly shaken, Western missionaries were preaching of a new god and code by which to live, and Western scientific discourses entered the land and caused many to rethink ideas that had long been held as indisputable. Along with the introduction of Western religions and science came the influence of values such as the right for love-based marriages and self-realization. The advent of Western science and Christianity was to have the greatest impact in creating negative stereotypes of gays in modern Korea.

The writings of some early missionaries to Korea demonstrate a strong contempt for what they understood as immoral behaviors. The missionary George Herber Jones writes of the Buddhist monks: "They debauched the people, and their abominations beggar description. The monasteries became pleasure houses, and the nunneries little better than brothels" (Jones 1907, 61). While perhaps not an outright condemnation of same-sex relations, the contempt of Westerners for things "not Christian" is highly visible in their writings. While not a direct criticism of homosexual activities, we can understand the contempt that a westerner such as the late-nineteenth-century traveler Isabella Bird Bishop ([1898] 1970, 56, 143) held for those of the eunuch class, whom she criticizes for their "very baneful influence on Korean affairs." She is more direct in her criticism of monks, who she claims were well-known among all people for their "gross profligacy." Following much the same pattern as in the West, scientific and Christian discourses changed the manner in which human sexuality was understood in Korea. Further compounding this hostile environment toward gays were the still-important remnants of Confucianism in Korea that stressed the need to continue the family lineage through male heirs, thereby fulfilling the inherent debt that all owe to their ancestors for being created.

The result of the influx of Western thought and religion thus helped shape the present-day view of gays in South Korea. The scientific analysis and categorization of homosexuals as an aberration from the norm, a sickness to be examined and treated, became accepted. And Christians, as Tamsin Spargo writes in her insightful examination of Foucault and queer theory, "developed universal moral codes and interdictions increasingly centered on the *truth of sex,*" which allowed homosexuality to be viewed as "intrinsically evil" (1999, 25). Korean Christian discourse on homosexuality is no different from that in the West. Contemporary society, then, can be characterized by a clear inability to separate sexual preference from gender.[6] That is, any actions beyond what is prescribed as normal (i.e., heterosexual sex) are not only immoral but also unnatural and unhealthy. These Western scientific and religious understandings of homosexuality combined in Korea with the neo-Confucian ethic of filial piety to form today's hostile environment for gays in South Korea.

Still yet another component that has added to the troubles for gays is the perceived breakdown of "traditional" family values by some Koreans. Various elements in the process of modernization, such as urbanization, have led to the disappearance of the extended family household that figured prominently in late Chosŏn Korea. While economics and individual desires have certainly contributed to this situation, the recent prominence of gays has also added fuel to the perception of a loss of long-standing traditions, regardless of whether either the tradition of the extended family or its breakdown is accurate. The fact that gays are on a more visible stage than in the past has led to views that their increased numerical growth is either an outgrowth of the breakdown of the family or a contributing cause.

Contemporary Discourse on Homosexuality in Korean Media

We do find a few references to gays in contemporary Korean literature, although the spaces in which gays are depicted are often heavily stereotyped or imbued with negativity. For example, in Yi Ch'ŏng-jun's short story "Pyŏngsin kwa mŏjŏri" (Fool and idiot, 1967), a deranged and psychopathic platoon sergeant terrorizes the men in his charge with the threat of death if they deny his sexual advances. Different from the other recruits in the story is the one described as having a "pretty face [and being] femininecontoured and a little chubby" who became the favorite of the sergeant and even seemed to enjoy his abuses. Yi depicts both of these men as outcasts in his story and the protagonist ultimately kills—albeit metaphorically—the sergeant when faced with the prospect of fleeing together with him. The author's image is decidedly an unsettling one that confirms the abnormality of homosexuals.

The short story "Hanak'o nŭn ŏpta: Sokssagim, sokssagim" (There is no Hanak'o: Whisper, whisper), written by Ch'oe Yun (1994), describes a woman through the eyes of a group of young men who are unable to see anything more in her than an occasional friend or pen-pal. Through glimpses of the woman in the story, however, the reader can see her successful business life and also the strong personal bond she shares with her constant female companion. While Ch'oe does not truly focus on her female character in this narrative, there are numerous hints at her lifestyle and the depth of her relationship with her partner throughout. For example, the conclusion of the narrative describes the two women as being "alternately business partners and companions." Also notable is that these two women live in Italy rather than Korea, perhaps indicating that such a lifestyle is better suited to a country other than Korea.

A significant addition to literature by and for gays in South Korea was realized with the advent of the e-journal *Gay Literature No. 1: Companion Mook,* published by Gaymunhak.com in May 2001. In this journal, gays—in fact the entire LGBT community of South Korea—are able to find a source for information, well-written literature, and contacts with others who might be sharing the same hardships. The signature piece of the maiden journal was "K'ŏming aut" (Coming out), written by Chŏn Myŏng-an (2001). This short story, based on both the author's personal experiences and of those around him,[7] details the difficulties faced by one man after coming out to his family—a declaration that is never accepted by his staunchly Christian mother and perhaps representative of the hardships that others could expect. There have been numerous other releases by this publisher in subsequent years, demonstrating the growing market for this type of literature.

Recently there have been a number of films that have attempted to come to terms with the problems that gays might experience in their lives. Perhaps the best example is the movie *Pŏnji chŏmp'ŭ hada* (Bungee jumping), which describes the blossoming of a relationship between a high school teacher and his student (Kim Tae-sŭng 2001). Yet the possibility of "gay" love is somewhat tempered by the inclusion of a Buddhist karma theme, suggesting that a former female lover of the teacher has been reborn into the male body of the student. Moreover, the difficulties for gays in South Korea are brought into relief by the fact that the teacher is fired and the student ostracized by his classmates; also noteworthy is that the movie ends with the pair in New Zealand—implying, at least superficially, that gays may only find happiness elsewhere.

A more recent movie that also aroused controversy by broaching issues of gay sexuality is *Rodŭ mubi* (Road movie) (Kim In-sik 2002). In this film, the love relationship between the two male leads is not a dramatic device to enhance melodramatic impact as in *Pŏnji chŏmp'ŭ hada* but rather a part of the overall relationship between the men. Gay sex is thus seen along with friendship, compassion, and love, providing a fuller image of what life might be like for these characters. In this sense, the movie depicts gays as "whole" people and gay sex as simply a part of a "whole" relationship. Nonetheless, despite providing a fuller image of gays, the movie cannot be said to portray gays and gay space with complete accuracy, as it seems to exoticize the lives of its main characters.

Along with the increased profile of gays in literary works and film have been the acknowledgments by some entertainers of their own sexuality. While there have been some notable individuals who have "come out" in recent years, there is little doubt that homosexuality and LGBT persons are persecuted and oppressed in South Korea. The case of the television personality Hong Sŏkch'ŏn perhaps demonstrates the "tolerance" curve that many gays could expect to experience by publicly coming out. In 2000, Hong announced that he was gay and was promptly fired from his television roles, his career seemingly over. While Hong has stated that his sexual preference was well known in Korean television circles, he felt anguished at hiding his identity and thus made his public statement (*Times on the Web* 2003). Yet three years after his exile, Hong is back on television and has seemingly resurrected his career, albeit not at the same height as before. This might indicate that South Korean society is becoming more tolerant of gays.

Another instance that supports the idea that South Korean society is becoming more accepting of "alternative" sexual identities is that of Ha Risu, the stage name for Yi Kyŏngŭn. Ha underwent a sex change operation in 1998 and was finally allowed by a Korean court to be changed to a woman on her family registry in December 2002. This was after a series of mixed rulings by lower courts, some refusing to acknowledge the change of gender. Perhaps just as notable as Ha's victory in having her gender change officially recognized is her enduring popularity as one of South Korea's biggest pop stars.

Despite these seemingly successful individuals, there are still significant hardships for those who come out in South Korea. First, we can cite efforts by the South Korean government to thwart a burgeoning queer cybercommunity. In July 2001, the government began to systematically block access to gay Internet sites, labeling such sites as a "harmful media" to youth.[8] The courts also have yet to offer any real protection to gays in terms of partner benefits, discriminatory practices at the workplace, or parenting rights.

Other court rulings can be understood as anti–LGBT discrimination, such as the Supreme Court ruling against a female transsexual rape victim, stating that "sexual assault inflicted on a transsexual does not constitute rape" (Bahk 2002). Moreover, there are certainly social discriminations against openly gay individuals in many sectors of society. Homosexuality remains something that is seen as unhealthy and abnormal by many in Korean society today.

A paramount question concerning the discrimination toward gays is "Why?" Why should society concern itself with gays or their sexuality? The crucial problem lies in the challenge that a gay sexual identity provides to a heterosexual identity. Men and women require each other to provide the Other on which they construct their own sexual identities. The heterosexual model of one sex requires the opposite model for contrast and balance. Gays, naturally, create problems for plotting out a heterosexual identity, and here is a basic problem for South Korean gays. We can thus see how the personal site of the body and its boundaries evolves into a contested space where hegemonic social views of sexuality are interjected and even enforced at times.

"Safe Zones" in Contemporary Korea

Recollecting the notion that performance of an alternative sexuality is tantamount to making the body a site of resistance, we should examine how gays are able to positively lead their lives within a society that largely rejects nonhegemonic sexuality. David Bell and Gill Valentine (1995, 146–147), in their study of lesbian women, note that these women most often "adopt a mantle of heterosexually defined behavior to avoid conflicts" in public spaces such as the workplace. Yet these women also bear "lesbian signifiers" knowable to some other women. This, according to Bell and Valentine, allows them to maintain multiple identities in a single space or time by manipulating time-space strategies to segregate audiences with "geographical boundaries or spheres of different identities."

In much the same manner, the most common approach in the Korean gay community is that of using a public space–private space model. The public space is a stage where a nonchallenging, mainstream-acceptable identity is performed; the private space is the stage that Korean gays share with their friends. The degree of boundary pushing by an individual is his or her challenge to hegemonic notions of sexuality, and thus one's site of resistance can be expanded or contracted depending upon personal desire or empowerment. Such a strategy of multiple identities provides social safety.

Despite the discrimination that many gays might face in today's Korea, there are concrete spaces where gays and the larger LGBT community can pursue the lifestyle of their choice. These spaces are places where greater rein is given to the idea of expanding boundaries and challenging hegemonic notions of sexuality. It is not surprising that such spaces have developed in areas where other norms of Korean society seem to have either been fully or partially jettisoned. As discussed in Sallie Yea's essay in this volume, *kijich'on,* or the areas surrounding US military bases, have long been spaces where nonmainstream behaviors are tolerated, and this also extends to those pursuing same-sex relations.

Kijich'on are "foreign spaces" having been designated as special tourist areas by the Korean government. Thus, alongside various shops peddling clothes, music, and souvenirs, there are numerous establishments that cater to the sexual needs of the US military personnel and others. In the same manner that this foreign space allows the operation of bars and brothels that provide for the heterosexual needs of foreigners, it also allows the operation of bars and nightclubs catering to gays, for both Koreans and others.

The It'aewŏn area, located adjacent to the US Eighth Army headquarters in Seoul, has perhaps the largest concentration of gay bars in Korea. These clubs are mostly in the tradition of other It'aewŏn bars, featuring English-speaking servers and an atmosphere that caters to both Koreans and others. The dozens of clubs concentrated in this small area demonstrate that there is a sizable gay community in Seoul.[9] Moreover, that these clubs are thriving in space created by the *kijich'on* offers strong evidence that these areas are governed by a different set of social rules—one that allows space for otherwise proscribed activities.

Such "safe zones" have more recently expanded to areas beyond the *kijich'on.* Some areas frequented by college-age students, such as Sinch'on, now boast numerous LGBT clubs. The growth of such clubs in areas beyond the *kijich'on* demonstrates that some Koreans desire their own spaces to gather, without the added problems oftentimes associated with non-Koreans. Such a development also demonstrates a desire on the part of some members of the Korean LGBT community to expand the boundaries of the stage where the performance of different models of sexuality is allowed.

One of the first Korean-orientated clubs was Lesbos. Originally opened in Map'o in May 1996, this club for lesbians has since relocated to the Sinch'on area. The club's Web site notes that the mission of the club has always been to provide a space for comfort and companionship for those otherwise alienated by society.[10] The fact that LGBT–orientated clubs have established footholds outside of the *kijich'on* reveals that many individuals now feel sufficiently empowered to move beyond the safety once afforded by more-or-less foreign areas.

Cyberspace and Gay Space

Another sphere that offers a safe haven for gays and gay culture is the Internet. Not only are there e-books and journals such as *Gay munhak* (Gay literature), there are dozens of chat sites, bulletin boards, and self-help Web sites. Interaction on the Internet supplies a degree of anonymity and allows individuals to interact with others shielded by a certain level of protection that face-to-face encounters do not offer. This is particularly true with finding support, information, and comradeship.

We can understand both the safety and lifeline afforded by cyberspace in the narrative by the aforementioned Chŏn Myŏng-an. Chŏn's protagonist, being forced apart from his gay contacts by his family, seeks to find companionship and comfort on the Internet by cruising chat rooms and bulletin boards (Chŏn Myŏng-an 2001, 43). The narrator demonstrates that cyberspace is a place where he can connect with others who will understand his situation. Despite being away from his home, the comfort of a "community" is always nearby via cyberspace.

Cyberspace is also a much-needed information source for many gays in Korea. Given a social climate that tends to isolate and push to the margins those who are different, individuals struggling with their identity would find it difficult to locate others with whom to exchange information.[11] Such emotional and geographical isolation can result in serious depression, especially in the case of youths. While statistical data is difficult to come by for the specific case of South Korean gay youths, studies in the United States demonstrate that the suicide rate for this group is potentially three or four times higher than that of their presumably heterosexual peers.[12] It is for this very reason that the South Korean government's crackdown on gay cyberspaces is troubling, as it could deny what could be life-saving resources for young people.

Yet the very nature of cyberspace is one that defies regulation to a large degree. From personal computers, there is no easy means to block access to Web sites, and the South Korean government has no jurisdiction over Web sites operating on servers outside of Korean territory. Moreover, despite attempts to block various sites with filtering software, successful Web site designers and operators somehow manage to stay technologically one or two steps ahead of those attempting to control cyberspace.

The growth of cyberspaces established for gays in Korea began in earnest from 1999 through 2000. During this period, a number of LGBT–orientated Web sites were established and provided a much-needed space for information exchange. Leading sites included X-Zone (Eksŭ-chon) and the lesbian site TGNet (Tank Girl Net).[13] From a limited number of sites initially, by 2000 numerous gay sites were founded and provided a measure of empowerment to Korean gays. Such Web sites are now too numerous to cite; this popularity can be attributed to the important function these cyberspaces fill for information exchange in the LGBT community.

It could be posited that from the Internet, gays and gay culture will be able to work into mainstream South Korean society in the same manner that other aspects of the Internet—such as e-mail, chatting, and cyber-education—have become a part of daily life.[14] Also emerging from cyberspace is a growing sense of empowerment and political rights among some gays. At the lead of such a movement are groups like the Lesbian and Gay Human Rights Federation (LGHRF) (Tongsŏng aeja in'gwan yŏndae).[15] This group's Web site provides information and news relevant to gays, helps organize gatherings and demonstrations, and features bulletin boards where advertisements, announcements, and personal queries can be posted by visitors. Important to any burgeoning political awareness by members of the LGBT community is organization and dissemination of information, and such a need is well met by LGHRF and other such groups. Yet the ongoing campaign being waged by the LGHRF against the government requirement for those entering "adult" Web sites, including many LGBT sites, to register by supplying their name and citizen identification number demonstrates that there still remains a significant fear of government persecution among some gays.

Expanding Sites of Resistance

An important idea for the LGBT community in Korea is that of the "politics of transfiguration" (Said 1993). Such transfiguration necessitates acceptance of differences and breaking down barriers that confront everyone in daily life. Thus a social space resultant from this transfiguration "can express and encourage an openness of outlook based upon a freedom to move across borders and boundaries in pursuit of new senses of self and other" (Pile and Thrift 1995, 21). Any type of transfiguration concerning gay sexuality, however, requires that an increasing number of gays accept the notion of the body as a site of resistance and help engender change in the hegemonic social discourse concerning sexuality.

Yet there remain many obstacles for the LGBT community despite the relative progress made in the past several years. While there might be an increased level of societal acceptance of gays over the foreseeable future, there remain many spaces where gay identities will face resistance and continue to exist only on the margins of society.

Aside from political and economic issues, an immense challenge for many gays in South Korea is to break away from parental expectations reinforced by the Confucian concept of filial piety. The social pressures to marry and have children cannot be easily dismissed even in contemporary Korea, and these thereby compound the difficulties with which some gays are confronted. The combination of still-important Confucian ethics and twentieth-century prejudices has created a very hostile environment for gays, both at home and outside.

A critical battleground for altering the societal view of gays seems to center on dismissing the intolerances against gays and their sexual preferences. Judith Butler (1997, 7) has stated that there needs to be a clear distinction between "'sex' as anatomical identity and 'sex' as regime or practice." What is crucial for the acceptance of gays is simply a separation of what one does from the person him/herself. While it seems almost trite to state that people should simply be accepted for their actions, this seems the fundamental problem for those individuals who choose a lifestyle contrary to what is accepted as mainstream.

Further, we should remain cognizant of the social implications of being gay on a public stage in regard to opportunity, economic reward, and public perception. Nancy Abelmann (2002, 35–36) has written in reference to Korean women that there are "particular associations between gendered personal attributes and social mobility inclinations." These associations surface in various media and shape public understanding of "types" of people. Although Abelmann reminds us that these perceptions are dynamic and changeable, I believe we can envision that by publicly coming out, a Korean gay would severely limit the avenues open for personal upward mobility. The negative images fused with homosexuals would simply not permit most individuals to overcome the many resultant difficulties and restrictions.

Such factors have prevented the LGBT community in Korea from establishing a strong voice. Without such a strong voice, gays have little chance to become politically and socially empowered and able to bring about significant societal change. The extension of the boundaries of the body toward a more visible site of resistance seems a slow process that has yet to gain momentum. While the question of empowerment has many different answers depending upon the individual, the ideal of living one's life unfettered by the prejudices of others has a strong appeal to all humans. As of yet, however, the refusal to turn the Other into the Same has not become urgent enough to engender significant social change. The revolution for gays in South Korea, like that of many other social minorities, has a great distance to go before true change is realized.

Notes

1. I acknowledge the problems with the use of terminology such as "homosexual" and "sexual orientation." Edgar A. Gregerson (1996, 1,174–1,175) comments that "homosexual" has been assigned as a pathological medical term and that the use of the term "sexual orientation" is also criticized by some who see willful decisions at work, thus preferring "sexual preference." Yet words of some type are needed to

discuss gays in South Korea; thus I translate *tongsŏng aeja* as "homosexual" and *kei* as "gay."

2. The terms *"tongsŏng aeja"* or *"tongsŏng yŏnae"* are relatively recently created words coined to match the Western terms "homosexual" and "homosexual love."

3. While Foucault gives a specific date and work for inventing the concept of homosexuality, other scholars see the date as earlier (i.e., late seventeenth century). For a fuller discussion, see Jagose (1996, 10–16).

4. Korea is not alone in this aspect: Mark McLelland (2000) describes a very similar situation in premodern Japan. Sunagawa Hideki (2006a) describes same-sex sexual relations in premodern Japan as a "history of silence" and echoes the idea that these relations were simply not role-defining activities.

5. Yet the same work notes the fine line between homoeroticism and favoritism and that a good part of the criticism had much more to do with favoritism than homosexual sex.

6. Judith Butler (1999, 10) writes of the preconceived roles for men and women, "When the constructed status of gender is theorized as radically independent of sex, gender itself becomes a free-floating artifice, with the consequence that *man* and *masculine* might just as easily signify a female body as a male one, and *woman* and *feminine* a male body as easily as a female one."

7. Personal communication with Chŏn Myŏngan, November 27, 2001.

8. The Ministry of Information and Communications adopted an Internet content rating system classifying gay and lesbian Web sites as "harmful media" and enforced their blockage of such sites through the use of filtering software on all public computer facilities.

9. The Web site http://www.utopia-asia.com/seoubars.htm (accessed August 31, 2006) lists dozens of clubs in this area, also providing a commentary on each.

10. http://lesbos.jinbo.net/htm/main_all.htm (accessed August 31, 2006).

11. I again note a very similar situation in Japan of the lack of reliable information and support for gays. See Sunagawa Hideki (2006b).

12. There have been numerous surveys on the suicide rate for American gay youths. The Web site http://www.youth-suicide.com/gay-bisexual/news/studies.htm (accessed August 31, 2006) lists various studies.

13. TGNet (http://tgnet.co.kr/ [accessed August 31, 2006]) was founded on August 21, 1998, and was the first Korean site dedicated to lesbians.

14. Of course, increased visibility does not necessarily equate with acceptance. Mark McLelland (2000, 4) notes, in the case of Japan, that the increased visibility of gays has not created space for individuals to come out.

15. The group's Web site is http://outpridekorea.com (accessed August 31, 2006).

11

Marginality, Transgression, and Transnational Identity Negotiations in Korea's *Kijich'on*

Sallie Yea

Kijich'on are US military camp town areas in South Korea (hereafter Korea). They are "foreign spaces," rarely bridged by Koreans apart from those who live and work there or who are born there. By "foreign spaces," I refer to both the inhabitants of the *kijich'on*—who are primarily transnational labor migrants, including foreign female "entertainers" and US military personnel—and the physicality of the townships, which are more internationalized (specifically Americanized) than virtually any other towns or suburbs in Korea. In this chapter, I suggest that *kijich'on* offer opportunities for the migrants who dwell there to engage in transnational identity negotiations and forge "strategies of everyday life" reflecting their transnational location (Castles 2002). In this sense, *kijich'on* are simultaneously sites of marginality within the Korean nation-state and, for some of the foreigners within them, sites of transgression and possibility.

Discussion in this chapter is situated within current debates about transnational migrants and their engagements with "host" or destination societies. In this era of globalization (of which Korea claims to be a major player), migration is seen to be the defining experience. Current migration patterns and experiences in the global age, according to many (e.g., Appadurai 2001b; Castles and Miller 1998; Faist 2000), exhibit quite distinct characteristics from those of past eras. In particular, in the current age migrants are increasingly "orient[ing] their lives to two or more societies and develop[ing] transnational communities and consciousness" (Castles 2002, 1,146). Much of the literature on the current era of global migration focuses on the identity negotiations that accompany the formation of these transnational communities and what they might imply for national identity, citizenship, and social change for nation-states—both homelands and destinations. In particular, much of the current literature seems preoccupied with how (and how successfully) migrants are able to create identities in their destinations through creatively adapting (shedding, exaggerating, reifying, reinventing) their existing national and cultural identities to the new context. This has lead Appadurai (2001a), among others, to proclaim the significance of "grassroots globalisation," where migrants not only are products of global change whose movement is tied to the broader structural political-economic environment, but who themselves craft transnational futures by constructing and reconstructing identities and communities, asserting considerable agency as they do so.

In these discussions about migrant identity and transnationality, little critical attention has so far been directed toward what these new contexts (destinations, host societies, and so on) actually look like and how the precise location of migrants within specific areas of the destination country, as well as the reasons for their migration to these areas, will jointly impact on migrants' ideas of community, marginality, identity, and belonging. Friedman, in exploring the "geographics of identity" that underpin migrant experiences, is one of the few who identifies "the significance of space as a situational marker of identity" (1998, 23). It is, in other words, the primary context in which transnational identities are made and remade. This chapter takes up Friedman's call for attention to the "significance of space and place" for Korea's *kijich'on*. Specifically, much of the current literature assumes that migrants will attempt to engage constructively with their host societies. But many migrants, including those in Korea's *kijich'on,* enter spaces in destination countries that are already, in a fundamental sense, "foreign" (and therefore contained) and in which, indeed, participation in or negotiation with the destination society is unwelcome or undesirable. Thus the field of identity negotiations in Korea's *kijich'on* is markedly different from those in which migrants interact with and to varying degrees reshape the destination country. Yet *kijich'on* offer no less a powerful field for constructions and negotiations of migrant identity through everyday practices that invoke normality, routine, and belonging.

Kijich'on and their residents have only been the subject of limited academic attention to date (Moon 1997; Sturdevant and Stoltzfus 1992), with research generally examining the relationship between the US military and prostitution and *kijich'on* figuring as a backdrop or spatial referent to discussions rather than as a thematic concern.[1] The six largest *kijich'on* in Korea are America Town in Kunsan (North Chŏlla Province, outside Kunsan Air Base); Tongducheon (Kyunggi Province, outside Camp Casey); Uijongbu (Kyunggi Province, outside Camp Stanley); Songtan (Kyunggi Province, outside Osan Air Base); It'aewon (Seoul, outside Yongsan Army Base); and Pyungtaek (Kyunggi Province, outside K-6 Army Base). The presence of US military personnel and foreign tourists is far more pronounced in these areas than Korean nationals, and indeed some *kijich'on* areas have been designated as "Korean Special Tourist Districts" by the Korean Special Tourism Association (KSTA), thus demarcating these districts as "foreign areas." Tongducheon and Songtan, both areas with large numbers of US troops, became Special Tourism Districts in 1997. Almost all the members of the KSTA are club or bar owners in *kijich'on.*

Filipinas and Russians began to enter Korea as entertainers to work in the *kijich'on* clubs and bars in late 1995. In 2001, according to National Statistics Office figures, approximately 1,600 Filipinas and 3,500 Russian women entered Korea on E-6 visas, which together comprise 60 percent of the total entertainers that entered Korea for that year, reflecting a similar pattern to that of Japan.[2] Although it is difficult to ascertain precise numbers of Filipinas and Russian women working as hostesses/entertainers in the various *kijich'on,* based on numbers of official and unofficial entrants, as well as counts of women made in clubs, one may safely estimate that there are at least 5,000 at any one time. It appears that many more Russian women also work in Korean prostitution districts as well as *kijich'on.*[3]

Although this chapter's discussion relates to all Korea's *kijich'on,* my focus is specifically on Tongducheon, both because it holds a particularly exaggerated significance as a symbol of American violence and occupation in Korea and because it, more than other *kijich'on* areas, has recently been subject to renewed discursive attention as the epicenter of the "sex trafficking problem" in Korea.[4] This recent addition to the array of images that mark Tongducheon as marginal has perpetuated its containment and problematization in Korea.

Discussion in this essay is based on eighteen months of fieldwork in Korea. This fieldwork period involved extensive ethnography and participant observation primarily in Tongducheon, where I also lived on and off at the houses of three Filipinas who had run away from clubs there to marry their GI boyfriends. For reasons of personal safety and anonymity, it was difficult to live in any *kijich'on* area on a full-time basis.[5] After one year of fieldwork in Tongducheon, I began fieldwork in other *kijich'on* areas, including It'aewon, Songtan, Pyungtaek, and Kunsan, although my engagement with these other areas was much more limited than Tongducheon. In particular, I began spending more time in Songtan, during which I developed a particularly close relationship with six Filipinas who worked in one club there, and I carried out ethnographic work with these women. Songtan thus became my secondary research site; in fact, it assumed a more important role in my research in 2003. During the fieldwork, I interviewed approximately one hundred foreign women working in *kijich'on* clubs and had regular informal discussions with around fifteen US GIs stationed at Camp Casey (Tongducheon), Camp Red Cloud (Uijongbu), and Osan Air Base (Songtan), respectively.[6] As well as interviews and ethnography, this research also draws extensively on a content analysis of media articles that help construct the discourse that maps nation-states in Korea as marginal places in the popular Korean imaginary. The chapter first examines the tropes that figure in the construction of *kijich'on* in Korea, then how these tropes act to exclude and discriminate against Koreans living in these areas, and finally explores how those migrants dwelling in these areas negotiate (and attempt to transgress) their marginality.

Placing *Kijich'on:* Sexual Deviance and Prostitution

Prostitution and sexual deviance are always spatially and socially constructed. Considerable work exists on how prostitution and prostitutes are discursively constructed, focusing variously on the powerful images and stereotypes of diseased bodies (sexually transmitted diseases and HIV/AIDS) and drugs and the presence of other illicit activities, including organized crime. Spatially, these images mark red-light districts (and by implication, sex tourism districts) as diseased, dangerous, and deviant.

Bruce Cumings, who drew on notes in a personal diary that reflects on a visit to the *kijich'on* outside Camp Stanley in Uijongbu, describes these sociospatial anxieties:

> The areas near the American military base are drastically deprived . . . here the local parasitic population lives . . . [in] filthy, backward, shameful "living establishments." The worst aspect, though, is the whoring district. There is no district—this aspect permeates that side of the town. But there are clusters of "clubs," catering only to Americans. Rock-and-roll blares from them, they are raucously painted and titled, and ridiculous-looking painted Korean girls—often very young—peer from the doors. . . . Several of them hooted at me as I walked by . . . but the most disconcerting of all was a middle-aged woman with two kids hanging to her who, in the middle of the street, asked me to come and "hop on" in the *chimdeh* [bed]. The town is crawling with mixed blood children. They seem fairly well cared for—certainly not beggars. Several Negroes [were] playing with the other children. . . . These kids are probably well-cared for because the whole town is in the dregs, removed from the prejudices of wealthy Koreans [and] from the terrible treatment of mixed kids that results (Cumings 1992, 171).

Despite his "thick description" of *kijich'on,* Cumings' discussions of militarized prostitution in Korea do not recognize the mutually constitutive effects of place, identity, and containment that mark *kijich'on* geographies in Korea, although he notes the "terrible prejudice" to which Amerasian children who are the products of *kijich'on* liaisons would be exposed if they moved outside these areas. Cumings' account was written at a time in Korea before the large-scale appearance of foreign women in *kijich'on* clubs and the "diverse race market" (see Kim Dong-hun 2002 below), which has witnessed the influx of foreign factory and construction workers from South Asia, Russia, and Commonwealth of Independent States (CIS) nations Nigeria and Southeast Asia into *kijich'on* (as both residents and customers in the clubs).

Since the entry of foreign women into *kijich'on* (and Korean prostitution districts) in the late 1990s, the local and foreign media has largely taken over the discursive production of *kijich'on* according to the trope of "sex trafficking." Sex trafficking has, for the first time, placed *kijich'on* on an international stage and further situated these areas in the Korean imaginary as "problem areas" where vice, prostitution, and organized crime are endemic. With the onset of international media attention on Korea's sex trafficking problem, *kijich'on* have thus also been the cause of national embarrassment and shame.

The mapping of *kijich'on* according to the discourse of sex trafficking began in March 2002 when Fox Television broadcast an undercover investigative report documenting participation of US military forces stationed in Korea in the commercial sex industry. In the report, American Courtesy Patrol Officers willingly shared their experiences and pitched advice on the operation of the sex industry in *kijich'on* areas (the report was filmed exclusively in Tongducheon). It was clear from the report that US soldiers were not only aware of the existence of sex trafficking in *kijich'on* but that many actively participated in it or perpetuated it by protecting those personnel directly involved. Women in the *kijich'on* bars from Russia, CIS countries, and the Philippines were filmed stating that their passports had been taken away from them by their managers or club owners and that they could not move about freely or return to their home countries if they wished.

Following this report, a spate of similarly revealing international print media articles carried exposés of the links between sex trafficking and the US military, with Tongducheon and Songtan acting as the primary sites for the stories. The American military newspapers, the *Army Times* and the *Navy Times,* carried an article that featured the story of Lana, from the Kyrgyz Republic, who was placed in a club in Songtan where she was forced to engage in prostitution for American servicemen (McMichael 2002). She lived in a three-bedroom apartment with nine other women from the club with a video camera mounted on the entrance so that the owner could monitor the women's movements. Lana stated that they were only allowed out for thirty minutes a day. The article also refers to the *kijich'on* area outside Camp Hovey in Tongducheon, conferring the following description:

> About a mile southeast of Tongducheon just outside the U.S. Army's Camp Hovey in the northern reaches of South Korea, six stern, broad-shouldered

> military police wearing black armbands over their camouflaged shirtsleeves stood on the corner outside Olympus Club and at least a dozen others walked the streets nearby. Inside the dimly lit club, loud, slow rock music pounded the ears. The club's 18 or so women, dressed in black bikini tops and miniskirts slit up the sides, were scattered throughout the club; standing at the padded bar on the right, ordering drinks or checking the new arrivals; seductively slow dancing with off-duty troops on the dance floor; talking or making out with troops in the padded, high-backed booths; or lap-dancing, grinding suggestively as they straddled seated troops (McMichael 2002).

This image, like the Fox Television story, was broadcast to an international, primarily American audience. Similar stories soon emerged, with Tongducheon as the main spatial referent (Kirk 2002; McIntyre 2002). The local press soon followed suit, so that the Korean public began to reimagine *kijich'on* as the actual and symbolic location of the transnational crime of sex trafficking and national shame of massive human rights abuses (for example, *Korea Times* 2002).

Nationalism, Anti-Americanism, and the Violence of *Kijich'on*

Kijich'on have not only entered the popular Korean imaginary as a symbol of deviant social/sexual behavior and sex trafficking but also as a symbol of anti-Americanism—particularly of the supposed ongoing colonization of the Korean Peninsula by hegemonic foreign forces (the United States) and the actual physical location of crimes by American GIs against the Korean people. Korea's modern history is littered with incidents of violent episodes involving GIs as perpetrators of various crimes against Koreans (almost always Koreans living in *kijich'on* areas). These include various incidents of violent, sexually based offenses against *kijich'on* prostitutes, the most infamous of which was the brutal murder of sex worker Yoon Kum-Yi in October 1992 by US serviceman Kenneth Markle, stationed at Camp Casey, Tongducheon. Over the past few years there have also been several highly publicized cases in which Korean prostitutes have been murdered or abused by US servicemen. In May 1993, a fifty-year-old woman who operated a bar in Tongducheon was found unconscious after being beaten and kicked in the head and face by a US corporal who was apparently trying to rape her. On January 30, 1999, also in Tongducheon, a Korean woman's body was found strangled in her rented room in Bosan-dong. Also in 1999, an American sailor was suspected of killing his Korean wife and their adopted son in their home outside Yongsan Army Base in Seoul. Petty Officer First Class James W. Fuhrman was investigated and later arrested on suspicion of killing the two, dumping their bodies a mile from their home in Hannam Village and then setting fire to them.

Such incidents have been used by numerous social movement groups in Korea as an argument for the withdrawal of US troops from the peninsula, which they see as contributing to the inability of North and South Korea to successfully reunify. As a consequence of two such incidents in 1993, in October of that year twenty civic and religious groups inaugurated a movement for eradication of crimes by US troops in Korea. The movement found that US soldiers commit an average of around 2,200 crimes in Korea each year, and that many of these crimes are sexually based offenses against women. Further, the movement highlighted the fact that less than 1 percent of crimes committed by US soldiers in Korea are tried in Korean courts.[7] Since the movement was established the general public in Korea has become more attuned to these negative consequences of the US military presence in Korea. As incidents continue to occur since the formation of the movement, they are now subject to popular protests and intense media coverage. The issue of prostitution in *kijich'on* is invoked by these groups in broader feminist-nationalist and anti-imperialistic discourses of anti-Americanism in South Korea. Kim Hyun-sook (1998), for example, discusses the way nationalist and feminist movements in South Korea have drawn on the issue of Korean women working in *kijich'on* clubs to push a particular nationalist-feminist agenda, thus often marginalizing the women's ability to narrate their own stories and experiences. For Kim Hyun-sook, these women constitute an "allegory of the nation" in that they are viewed as symbolic of the ongoing colonization and occupation of the Korean nation by imperialist forces, embodied literally by the sexual violation of Korean women and violence against them by US troops. This powerful metaphor situates *kijich'on* prostitutes of Korean nationality within discourses of gender and Korean nationalism.

The most recent incident contributing to this growing expression of anti-Americanism in which *kijich'on* are the direct referents involved the death of two female Korean middle school students who were hit and killed instantly by a US armored vehicle on June 13, 2002. This incident also occurred in Tongducheon near Camp Casey. It resulted in numerous public protests and rallies in the following weeks, including one demonstration on July 30, 2002, in which about twenty student activists stormed Yongsan, home of one of the largest US military bases in Korea. The protests escalated in early December when the two servicemen involved were acquitted of charges of negligence leading to wrongful death.

Marginal Koreans of Tongducheon: The Dream of Lee Kang-suk

"Tongducheon is a symbol of the suffering in Korea's modern history and the Korean peninsula. I hope that Tongducheon will be valued as Kwangju, which strongly symbolises democracy" (Kim Dong-hun 2002).

Lee Kang-suk runs an English-language institute in Tongducheon. In a three-page article in *Han'gyŏreh sinmun,* he reflects on his experiences of being a native of Tongducheon (Kim Dong-hun 2002). In a nation where localized (particularly regional) identities have an important bearing on one's status, opportunities, and self-understanding (Yea 1994 and 2000a), being from a *kijich'on* area is fraught with deep-seated meanings and pervasive stereotypes that can lead to discrimination and marginalization. Lee was born and grew up in Bosan-dong, Tongducheon. Bosan-dong is the area directly adjacent to Camp Casey, the largest and arguably most "notorious" of the US military bases in Korea. Most of the residents of Bosan-dong have personal histories that are bound in some sense to the US military presence there. Living in Bosan-dong, for example, are Korean women working the bars and clubs, foreign women who have run away from these bars, GIs who have married Korean or foreign women and taken up residence there, Amerasian children, owners of *kijich'on* businesses, and more recently, unskilled foreign factory and construction workers. These histories of belonging are signified by the informal naming of Bosan-dong as "American Alley."

As a child, Lee helped his father collect aluminium cans in a wheelbarrow from the nearby US bases to make boilers. He recalled that he would go to the dumping areas (rubbish tips) and sift through the cans, still bearing the remnants of their contents of American goods like butter, ketchup, and baked beans. As he passed through Bosan-dong on his way home, the GIs would often kick the cans he had tied together in amusement. His father's house was known to the GIs as "can house," and the GIs would sometimes go there on weekends and take photographs of his home. As he collected cans when he was young, he tried to read the labels, and an interest in studying English was instilled. He became a KATUSA (Korean Augmenti to the United States Army Korea) and he eventually graduated from Korea University with a major in English.

Attempting to move outside Tongducheon and enter "mainstream" Korean society created tensions and anxieties for Lee, including his experience of enrolling in Korea University. It also produced problems for one of his English academy students, as he recalls in the article: "When he first introduced himself in college as being from Tongducheon everyone in the class looked at him in a strange manner. He knew they were thinking, 'How could he have come to college from such a bad area?' That was the kind of look they gave him." Then, not long ago, one of his female students met a boy through an arranged marriage meeting. "She didn't lack anything as a potential wife, but the boy's parents rejected her because she was from Tongducheon."

Attempting to transgress the received stereotypes of Koreans from Tongducheon—as uneducated men or loose women, for example—was a difficult endeavor for Lee, and for his student it proved futile, as her rejection from Korean society beyond Tongducheon was sanctioned by the parents of her potential marriage match. Lee also recalled a recent incident when he was watching a sitcom on television in which a man was speaking "konglish" (a mixture of English and Korean) and was asked by another man, "Why is your accent like a person from Tongducheon?" Like the southwestern Homan region (where, until recently, Homan people were portrayed in sitcoms as lower-class people speaking in "low language" *[pan mal]*), Tongducheon people are portrayed negatively by Korean popular media. For Lee, the images of an American city, drugs, violence, mixed races, and prostitutes catering to foreigners coalesce to produce a powerful metaphor of Tongducheon as a deviant and evil place that is in Korea but not "Korean."

The broader significance of Lee's account in *Hanguryre sinmun* lies not only in his reflections on the experience of being a Tongducheon person but also his symbolic identification of Tongducheon with Kwangju city. Kwangju, capital of the southwestern province of South Chŏlla, has in recent years undertaken a self-promotion as a "Mecca for Democracy" (Yea 2002). Kwangju and South Chŏlla Province have managed to transform the negative connotations of images of radicalism, equality, and democracy/social justice into a basis for regional place promotion that spans both touristic and political arenas. Lee's desire for a similar transformation of Tongducheon is an appeal to recognition of the spirit of Tongducheon as a symbol of both historical suffering (at the hands of the United States) and the resilience of the Korean spirit.

The Women Outside

The mainstay of *kijich'on* economic and social life revolves around prostitution and informal sexual and romantic liaisons between prostitutes/hostesses and American GIs (and more recently, with the introduction of Filipinas and Russian women working in the clubs and bars, of unskilled migrant workers and Korean men). The institution of militarized prostitution in *kijich'on* has until very recently almost exclusively relied on the sexual labor of Korean women who both live and work in *kijich'on*. In 1999, some 27,000 women were reportedly employed in the various *kijich'on* throughout Korea, with average earnings of only around US $250 per month (Takagi and Park 1995). This amount was seen to be barely sufficient to cover food and rent, and many of the women were subsequently forced into substandard housing and an extremely volatile standard of living.

As well as their economic marginalization, the Korean women working in *kijich'on* continue to face extraordinary discrimination from other Koreans (this, ironically, despite the fact that their "cause" has been appropriated by various social movement groups in Korea, as discussed by Kim Hyun-sook 1998), and they are normally unable to reenter mainstream Korean society if they attempt to leave *kijich'on.* The documentary film *The Women Outside* tells the story of one young Korean woman, Yang Hyang Kim, who applies for a job in what she thought was a coffee house, only to find herself being sold to a brothel outside Camp Stanley in Uijongbu, Kyunggi Province (Takagi and Park 1995). When Kim finally succeeds in escaping from the brothel, she tells how she was rejected by both her family and society. She eventually returns to the *kijich'on* and continues working there until she is well into her fifties. These older Korean women are referred to by the term *"hiphari"* in Korean. Turyebang (My Sister's Place)—a nongovernment organization located outside Camp Stanley, Uijongbu, Kyunggi Province that addresses issues of Korean women in the *kijich'on* areas—claims that a major problem for these older Korean women is that they are not eligible for welfare until they turn sixty. However, for most of them a working life in prostitution ends when they are 45–50 years old, leaving a period of ten to fifteen years between the end of their working life and the time they can receive welfare in which they have no ready employment options.

Most of these women and their offspring find Korea beyond *kijich'on* extraordinarily difficult to negotiate. When they become too old or "damaged" to continue working in prostitution, they remain in *kijich'on* in various other capacities, including as *manasans* (bar managers or overseers) at the clubs where they may have worked formerly, owners of mom-and-pop stores or small restaurants, or as unemployed beneficiaries of workshops and alternative employment schemes run by local nongovernment organizations situated in *kijich'on* areas. Despite the appropriation of their voices by radical social movements in Korea, their exclusion from mainstream Korean society marks their difference and deviance in the same way as the other (foreign) residents of *kijich'on.*

Containing *Kijich'on*

As a result of the coalescence of images of prostitution, sexual deviance, and sex trafficking on the one hand and violence and anti-Americanism on the other, Koreans and some long-term foreign residents of Korea rarely enter *kijich'on* areas. Those few Koreans who do visit *kijich'on* (apart from Korean men who go to engage the sexual services of the foreign women deployed in the clubs) tend to hold fears of bodily pollution and physical and moral danger through potential encounters there. One female Korean reporter who was directed to research a story on sex trafficking in Tongducheon recalled, "As soon as I came back home I closed the door and took all my clothes off and went straight to the shower. That place made me feel so dirty" (personal communication August 2002).

Another Korean friend explained, "If you are a Korean woman and you are from Tongducheon, everyone thinks you are a prostitute." When I asked her if it's the same if you are from, say, Uijongbu or Songtan, she replied, "Not really—maybe a little bit from Uijongbu, but for sure it's the image of Tongducheon. I try not to go there because I don't want people to think I am some kind of prostitute" (personal communication April 2003). Even long-term foreign residents in Korea regard Tongducheon contemptuously according to such stereotypes of prostitution and US military excesses. One American professor at a prominent university in Seoul, for example, stated "That place [Tongducheon] is the pit of hell. The GIs are especially bad out there—they should all be locked up on the base and not let out. They're like animals" (personal communication May 2002).

These same images, fears, and biases are also common to sex tourism districts. Discussions of sex tourism in academic literature often identify sex tourists and prostitutes as "dangerous forces; forces that, while subordinate to the mainstream of society, by their very presence challenge the norms of the dominant. Their existence continues to represent alternative lifestyles" (Ryan and Hall 2001, 2). Because of their "alternative" character, the geographies in which these relationships (between sex tourists and prostitutes) are expressed are contained and bounded. They become, in other words, the spaces and places of marginal people and activities, marked by relationships, encounters, and liaisons of deviance from societal norms and values.

While the boundaries and exclusions that mark *kijich'on* in Korea are not visible and physical, their containment to specific areas and their location within the sex tourism realm is the subject of governmental intervention, which fuses legality and legitimacy with marginality and difference. This intervention came in 1997 with the designation of two of the biggest *kijich'on* areas, Tongducheon and Songtan, as Korean Special Tourism Districts. Although the descriptions that accompany the official marking and promotion of these districts suggest shopping and bars and clubs as the mainstays of tourism, sex tourism is an implied element of the touristic experience in *kijich'on* areas.

Kijich'on areas are thus contained within the Korean nation-state through the establishment of Special Tourism Districts where only foreigners are supposedly allowed to enter and through the fears held by Koreans about the polluting or dangerous effects of entering these areas. Yet keeping the foreign residents contained within *kijich'on* is equally as important as Koreans remaining outside them. The movement of US military personnel outside *kijich'on* areas is tightly controlled and often subject to severe restrictions during periods of heightened anti-Americanism in Korea. After the death of the two middle school students in 2002, the incidences of confrontations between Koreans and GIs elevated. The United States Forces Korea (USFK) began placing severe restrictions on GI movement off base, introducing new curfews and placing several places off limits to GIs. On September 15 of that year, for example, a GI was briefly abducted by an angry mob and forced to make apologies in a university stadium (Kyunghee University, Seoul) after a scuffle broke out on the subway line to Uijongbu (the most frequently traveled route by GIs based north of Seoul). The incident occurred when the GI refused to accept a leaflet that was being handed out about the death of the two middle school students. In fact, the leaflet was in Korean, a language which the GI could not read. This incident demonstrates the possible consequences of movement off the bases and outside the contained spaces of *kijich'on* for GIs, particularly during periods of heightened anti-Americanism. Concern for the safety of the soldiers and the desire to minimize incidents such as the one that occurred on the train to Uijongbu have also led to the introduction of restrictions on soldiers' movement outside the base where they are stationed by the USFK.

The movement of the foreign female "entertainers" in Korea's *kijich'on* is even more highly restricted than that of the GIs by control and surveillance mechanisms utilized by the Korean club owners and managers. These include video cameras on apartment entrances monitoring the movement of women in and out of their residences, restrictions on free time during nonworking hours (such as being allowed out for only thirty minutes a day, or not at all as a form of punishment), and constant checking of women during their free time by calling them on their cell phones (Yea 2004a). In addition, women are told (incorrectly) by club owners that they are not allowed out of the clubs for the first three months they are in Korea, since they are not yet legal (having not yet secured their Alien Registration Cards). The women also face the constraints of being generally unable to speak or read any Korean language, having no idea of their actual physical location in Korea (and therefore how to get to other places, including Seoul), and they do not know the local transportation system in order to facilitate their physical movement outside *kijich'on.*

In addition to these practical/physical constraints, women who have already run away from the clubs to marry a customer or seek alternative work in a factory face the dual problems of illegal status in Korea (and therefore the possibility of being caught by the police or immigration authorities should they venture too far) and severe discrimination by Koreans. In September 2002, one of my research participants, "Valerie" (age eighteen, Filipina), recalled her disappointment at an outing with her GI boyfriend the previous day. She had gone hiking with her boyfriend to a nearby mountain and then to Seoul to watch a movie. It was her first time to go out of the apartment in three weeks since she escaped from the club where she had been deployed. She said she was so tired because "I don't go anywhere for a year," and "I don't like that. Koreans just look at me then talk to each other and laugh."

Kijich'on Migrants Negotiating Identity: Romancing the Club

This chapter has so far examined the constitution and perception of *kijich'on* and popular views of Koreans who live in or are born to these areas by the "Korean nation." But what of the "liminal or marginal people" (Ryan and Hall 2001, 3) who dwell in and are contained by *kijich'on?* What possibilities do *kijich'on* offer for transgression of their marginality and negotiation of their identity as undesirable transnational migrants in Korea? What kinds of translocal and everyday practices are forming in these spaces and how, if at all, do these challenge our understanding of migrant geographies? The discussion here follows Castles' observation for transnational migrants that these "groups and individuals find creative ways of simultaneously adapting to and changing their social environment. The human agency they develop applies not only to overt political or social action, but also to the strategies of everyday life" (2002, 1,158). Strategies to retain, regain, or create a semblance of everyday life are pursued by both foreign female entertainers and American soldiers in and through spaces of the *kijich'on.* Because *kijich'on* are oriented primarily toward fulfilling the sexual needs of the GIs in the surrounding American bases, the clubs, bars, and restaurants provide the primary spatial context in which these strategies and negotiations are played out.

In my discussion in this part of the chapter, it is not my intention to overly sentimentalize the experiences of the foreign women deployed in the *kijich'on* clubs. Nor do I wish to suggest that forming relationships with GI customers that transcend sexual labor somehow erases the power dynamics that exist in these relationships or the women's status as trafficked entertainers. There is no doubt that all of the women who participated in this research experienced extreme human rights abuses and emotional trauma while deployed in Korea and that some of these abuses at least occurred at the hands of GIs and other customers. It is simply that such abuses do not capture the complexity of these women's experiences, the agency that the women attempt to assert in negotiating their everyday lives, relationships and status as "trafficked entertainers," or the self-understanding and imaginings about the futures these women are constantly attempting to shape while in Korea. As I note elsewhere:

> Establishing romantic relationships with customers is one means by which these women are able to assert a degree of agency and work towards a (partial) reversal of the power dynamics between themselves, their employers and their customers and to overcome the stigma attached to their labels as "trafficked entertainers" or "prostitutes" by introducing the notions of romance, love and friendship into such encounters. These relationships usually begin as a source of amusement or fun and as a means to counter

the boredom of otherwise mundane lives in the *kijich'on*. In most cases GI boyfriends also play an important role in providing financial and emotional support for the women (Yea 2006).

Cohen's (1982, 1986) studies of relationships between Thai prostitutes and foreign tourists *(farangs)* in sex tourism districts in Bangkok provides an insight into the establishment and maintenance of romantic relationships between Asian women and foreign men in a similar context to that of Korea's *kijich'on*. Cohen (1982) found that relationships between prostitutes and foreign sex tourists in Bangkok came to hover on "the edge of ambiguity," with many of the elements of prostitution, such as barter, promiscuity, and emotional indifference, partly or completely absent. Heather Montgomery found that child sex workers in Thailand also construct their relationships with foreign customers in "words that conveyed an ambiguity and a conceptual difference between selling sex and working" (2001, 145). Such ambiguity characterizes many entertainer–GI interactions in the *kijich'on* clubs and bars of Korea, with women construing encounters with some customers in benign and even positive terms.

For the foreign women in Korea's *kijich'on,* the words describing their work are carefully chosen: Bar fines (see below) are often called "dates," and the women are often quick to point out that such dates rarely entail sexual intercourse. As Cheng points out, "romantic love" is an emotional discourse in the *kijich'on* clubs for both Filipinas and Gis, who are "constantly engaged in defending themselves against the label of prostitution" (2002, 103). Thus, because these Filipinas are normally trafficked—meaning they are coerced into performing sexual labor—their situations create a response to their work in the clubs among both themselves and their customers that actively counters perceptions of their liaisons as sex work or prostitution. In other words, because the women are not voluntary sex workers, they seek to construct their work via an alternative discourse of love and romance. Yet this discourse serves not only to neutralize the negativity that surrounds their lives in the *kijich'on,* it also actively constructs alternatives, including boyfriends, love, and sometimes even marriage (Yea 2004b).

Apart from the women's desire to negotiate and counter their status as entertainers in Korea, a combination of the women's vulnerable situations in *kijich'on* clubs, the system of ladies' drinks that operates in the clubs, and the women's constructions and comparisons between GIs and customers of other nationalities (Koreans and South/Southeast Asian unskilled migrant workers) helps bring GIs and the women together in romantic relationships. The drink system in the *kijich'on* clubs operates in the following way: Because the women normally receive less salary than has been agreed to in their contracts (or salary is withheld completely), in order to have a constant source of cash income they must generate money through a drink sales system and, for some women, by engaging in the provision of sexual services at the clubs. The drink system entails having customers buy them drinks for which they receive a percentage from the club. A drink for a woman normally costs between US $10 and $20 and, as an unwritten rule, each drink allows the customer to spend between fifteen and twenty-five minutes with the woman, after which the customer must buy her another drink. From the purchase of drinks, the women get between 1,000 and 2,000 won, or between 10 and 20 percent of the money, the rest going to the club owner. A book is usually kept behind the bar in which a record of how many drinks each woman sells in a week/month is kept.

It is this drink system that supplies the nickname GIs give to the women who work in the clubs, dubbed "juicy girls," "juices," or "drinkie girls." The name "juicy girl" derives from the fact that the usual drink for the women is a tiny glass of juice. Buying a "ladies' drink" is a symbolic act in that the drink represents a period of time for which a customer buys a woman's company or she "sells her personality" (cf. Hochschild 1983). What the woman is expected to do in the fifteen to twenty-five minutes of time she will spend with her customer for each drink will usually depend on the price of the drink, whether the woman works in a "good" or "bad" club, and what the GI expects.

The bar fine system can offer a similar space for romantic relationships to develop as the drink system, although GIs' initial encounters with women are usually through the drink system. A bar fine can be described as the purchase of a woman's time to take her outside the club. The length of time varies, and there is usually a set price for different periods of time, such as three to four hours or overnight. Again, as with the drink system, what a woman is expected to do with the customer on a bar fine depends on the club, the customer himself, and his negotiations with the woman. If the woman has already established a relationship with a customer (through, for example, repeated encounters in the context of the drink system), a bar fine is viewed in far more benign terms than if a customer purchases the woman's time without establishing some preceding relationship with her. Some customers—especially those that already "know" the women—view bar fines more like "dates" on which they take the woman barhopping, to a restaurant, or shopping. In these cases, even if the woman spends the night with the customer, there is not necessarily any expectation on his or her part that they will have sex. The distinctions between customers that are developed by the women through the drink system are thus perpetuated in the context of bar fines.

In sum, romantic relationships between foreign female entertainers and customers can develop in the space of the *kijich'on* clubs. Given the restrictions on women's freedom of movement and free time in many of the clubs, the drink and bar fine systems that operate in the clubs offer the only real space for relationships to develop. The drink system and sometimes the bar fine system enable the foreign women and GIs to talk, dance, and "have fun" in the club and go on "dates." However, not all GI customers view their encounters with women in such benign terms, and many customers simply wish to purchase a woman's time for her sexual labor. Such constructions, played out through the drink and bar fine systems, help situate *some* GIs as potential partners through a comparison with other GIs who are perceived as dishonest, disgusting, or violent. These distinctions are not just limited to different GI customers but also to customers of different nationalities.[8]

While many of the women who participated in my research did not necessarily aim to marry a GI (but see Yea 2004b), most viewed a GI boyfriend as a positive investment in their immediate future within the context of the work in the clubs and life in the *kijich'on.* For the foreign women who live and work in Korea's *kijich'on,* then, the *kijich'on* is their world. Although marred by the exploitative and abusive conditions of clubs involved in sex trafficking in Korea, the space of the club and the *kijich'on* strip offer women possibilities to simultaneously assert agency in forging alterative, postclub futures and creating strategies for everyday resistance within them (Scott 1985). For them, the clubs are places of both oppression and opportunity. The opportunity to meet GIs and form romances and possible marriage works to negate the oppression of the clubs while remaining in a transnational space.

As a result of relationships that develop through encounters with customers, as well as the generally deplorable working conditions in the clubs, scores of Filipinas and Russian women run away, hiding from their former employers and Korean police and immigration authorities in nearby private residences rented by GIs, Filipina or Russian friends, or in factories. The process of running away and marrying has in fact become relatively routinized, as women follow the path and seek the advice laid down by those who have run away before them. Consequently, in *kijich'on* areas surrounding the larger US military and air force bases in Korea, reasonably stable communities of interracial couples, including their children, have recently begun to form.

One of the largest of these communities is adjacent to Camp Casey in Tongducheon in the neighborhood of American Alley. At least 150 Filipina–GI couples (and the equally numerous population of Russian–GI couples) live here, along with fluctuating numbers of other women who run away from the clubs and "hide out" in the area with friends until a suitable husband or factory job can be found. This part of Tongducheon—about ten square blocks and consisting of a series of small alleys and back streets fronting poorly constructed, dilapidated single or two-story apartments—provides cheap accommodation and inconspicuous hideouts. Mapping *kijich'on* according to these communities further suggests the potential of marginal places in Korea to be transgressed and re-created as spaces of hope (cf. Harvey 2000).

Conclusion

Discussions of migrants and their geographies in Korea have been extremely limited to date, with studies tending to focus only on issues relating to challenges to Korean identity (Moon, Katharine 2000) and migrant human rights and labor conditions (Lee Hye-kyung 1997). Despite the large number of both legal and illegal unskilled labor migrants in Korea (approximately 300,000 at present, including at least 5,000 foreign women who enter on "entertainers" visas) and US military personnel (approximately 37,000 permanently deployed personnel at any one time), no research has yet focused on experiences of migrant identity and community among and between these groups. Certainly little research has appeared that critically examines the geographies through which these identity negotiations literally take place.

This chapter has described the geographies of *kijich'on* areas and explored the meanings attached to them by both Koreans and foreigners, including the US soldiers and the foreign women deployed as entertainers/hostesses in *kijich'on* clubs. The popular construction of *kijich'on* by Koreans (and some foreigners) focuses on a coalescence of images and tropes that define these areas as Other. These include sexual deviance, violence, crime, drugs, and associations with some of the worst vices of foreign (particularly American) culture. The connotations of these images have been perpetuated recently by the national and international media associations of *kijich'on* with the problem of sex trafficking. As a result of these images, *kijich'on* are constituted as marginal and foreign (i.e., inhabited by foreigners and largely off limits to Koreans) spaces in the popular Korean imaginary. Consequently, the foreigners (and Koreans) who live and work in *kijich'on* are subject to gross stereotypes and face extraordinary prejudice in "normal" Korean society. Movement outside *kijich'on* for those who live there is fraught with anxieties and the potential for violence, discrimination, or ridicule so that residents remain largely contained within their figurative and real boundaries.

In this chapter, I have focused on *kijich'on* as places in and through which migrant identity negotiations occur in Korea. While *kijich'on* and those who dwell there are marginal spaces within the Korean popular imaginary, they simultaneously (and consequently) act as fields for continual negotiation of identities of these transnationally situated migrant subjects, and they therefore offer possibilities for the reconstruction of identities and—for some—transgression of marginality. As the club romances between the American soldiers and foreign female entertainers in *kijich'on* clubs reveal, these spaces also offer residents the possibility of transgressing their marginal status through "strategies of everyday life" that enable these subjects to create, retain, and regain the social relationships that mark "normal" everyday life, particularly romantic relationships and partnerships. For these subjects, transgressing their marginality involves crafting socially sanctioned (for both the foreign women and the American soldiers) relationships that represent the possibility, if not the actuality, of a normal romantic relationship and, in some cases, the possibility of marriage.

Space and place are not at all insignificant to these transgressive identity negotiations that occur in and through the space of the *kijich'on* clubs and restaurants of the bar strip and the apartments and houses adjacent to the club areas, such as American Alley in Tongducheon. For foreign women in Tongducheon, as well as Korea's other *kijich'on,* the possibility of transcending their status as marginal people (for example, "sex slaves," "prostitutes," or "trafficked women") therefore exists in the same places that their marginality, exclusion, and victimization are expressed. Many are able to construct alternative futures for themselves through the possibilities that *kijich'on* can provide. Reading *kijich'on* this way opens up a new engagement with space/place in Korea by focusing on migrant spaces and how they are constituted and experienced by those who dwell in them and how migrant Others work to negate the marginality in particular and specific destination contexts.

Notes

1. The only study dedicated to an overview of life for the Korean women who work in *kijich'on* in Korea is the excellent chapter in Sturdevant and Stoltzfus' (1992) book on militarized prostitution in Asia. This chapter takes Tongducheon as its referent and draws on the narratives of two Korean women who live in Tongducheon. This study is thus unique in that it is the only one that centers discussion on the voices of the women themselves (see section, "The Women Outside").

2. See Ballescas (1992) and Matsui (1995) for background discussions on Filipina entertainers in Japan. Santos (2002) estimates that in 1999 there were 42,998 Filipinas employed as overseas contract workers in Japan in 1999 and that 99.6 percent of these women's specified jobs were entertainer.

3. One of the main reasons for this pattern of migrant geography is the preference US GIs tend to assert for Filipinas, since their English is normally much better than that of the Russian women. In the absence of comprehensive research on the situations of Russian entertainers, much of the information surrounding their experiences to date is based on speculation and casual observations.

4. Approximately one-third of the total land area in Tongducheon is occupied by US military bases and installations.

5. Because these are small, contained areas, the presence of a foreign woman like myself who was clearly neither a female GI or club worker would raise unwanted attention and suspicion. Weekly movement in and out of the *kijich'on* was a strategy aimed at maintaining my anonymity in this environment.

6. Formal research with US military personnel in Korea or elsewhere requires formal permission from the US Army Command in Korea. Given the nature and sensitivity of my research, which was oriented primarily to sex trafficking in Korea, I decided not to formally pursue this line of research. In addition, I felt that GIs were normally willing to discuss their relationships with women in the clubs and their status in Korea with me on an informal basis. Formal interviews through a sanctioned research process would, in my opinion, have constrained the research and not encouraged GIs to talk openly about their experiences.

7. Under the provisions of the ROK–US SOFA (Status of Forces Agreement), which governs the provisions of the US military in Korea, virtually all crimes committed by American military personnel in Korea are tried in an American military court, and defendants are placed in US military rather than Korean custody.

8. Although the clubs in *kijich'on* are established primarily for the entertainment needs of the GIs, Koreans and migrant factory workers also patronize the clubs. Club "B," where Julie and Emily (both 26) worked, is similar to many of the others in that it generally receives only GI customers, but if there is a lockdown (meaning the GIs are restricted to base), or if there are no GI customers, the *manasan* (club manager/overseer) will let customers from other nationalities enter. Migrant workers and Koreans are normally not allowed in the clubs until the GIs have returned to base to meet their nightly curfew, in part to avoid fights from breaking out in the clubs. All the women interviewed for this research preferred American GI customers to other customers. This preference was constructed as a result of a number of factors, including linguistic affinity (English was the medium for conversation) between the Filipinas and GIs, the regularity with which GIs were able to visit the clubs because of the proximity of the bases to the clubs (as opposed to migrant workers who, because of constraints of time, money, and proximity, could often visit the clubs only on the weekends), and the benevolence expressed by some of the GIs toward the women. This attitude (discussed in detail in the following section) was often demonstrated when GIs would give women money or buy them food and personal items—even if the GI was not the boyfriend of one of the girls. In addition, at a more fundamental level, when Koreans or migrant workers (although not Filipino migrant workers) went to the clubs, they would usually do so only for the purpose of sexual gratification (like some GI customers), whereas, as suggested above, many GIs would go to the clubs simply to socialize.

Works Cited

Abelmann, Nancy. 1996. *Echoes of the past, epics of dissent: A South Korean social movement.* Berkeley: University of California Press.

———. 2002. Women, mobility, and desire: Narrating class and gender in South Korea. In *Under construction: The gendering of modernity, class, and consumption in the Republic of Korea,* ed. Laurel Kendall, 25–53. Honolulu: University of Hawai'i Press.

Ahn Ch'ang-hyun. 1998. Kim Mo-im changkwan nosukcha hyŏnjang 'chamhaeng' sŏsomun kŭpsikso ch'aja'seoul-yŏk sŏ yasik to nanuŏjuŏ. *Han'gyŏreh sinmun,* Social section, June 3: 22.

Anderson, Benedict. 1983/1991. *Imagined communities: Reflections on the origin and spread of nationalism.* London and New York: Verso.

Appadurai, Arjun. 1996. *Modernity at large: Cultural dimensions of globalization.* Minneapolis: University of Minnesota Press.

———. 2001a. Grassroots globalization and the research imagination. In *Globalization,* ed. Arjun Appadurai, 41–55. Durham, NC: Duke University Press.

———, ed. 2001b. *Globalization.* Durham, NC: Duke University Press.

Arluke, Arnold B. 1988. Sacrificial symbolism in animal experimentation: Object or pet? *Anthrozoos* 2(2): 98–117.

Atkinson, David, and Denis Cosgrove. 1998. Urban rhetoric and embodied identities: City, nation, and empire at the Vittorio Emanuele II monument in Rome 1870–1945. *Annals of the Association of American Geographers* 88(1): 28–49.

Bae, Changmii. 2002. The symbolic landscape of national identity: Planning, politics, and culture in South Korea. Ph.D. diss. University of Southern California at Los Angeles.

Bahk, Sarom. 2002. Viewing the invisible minority: The state of LGBT persons in Korea. *Features e.Peak,* 2004, from Web site http://www.peak.sfu.ca/the-peak/2002-3/issue2/fe-klgbt.html.

Baldwin, Frank. 1979. Participatory anti-imperialism: The 1919 independence movement. *Journal of Korean Studies* 1: 123–162.

Ballescas, Ma Rosario Piquero. 1992. *Filipino entertainers in Japan: An introduction.* Quezon City: Foundation for Nationalist Studies.

Basso, Keith H. 1996. Wisdom sits in places: Landscape and language among the Western Apache. Albuquerque: University of New Mexico Press.

Bell, David, and Gill Valentine. 1995. The sexed self: Strategies of performance, sites of resistance. In *Mapping the subject: Geographies of cultural transformation,* ed. Steve Pile and Nigel J. Thrift, 143–157. London: Routledge.

Bender, Barbara. 1993a. Introduction: Landscape—meaning and action. In *Landscape: Politics and perspcectives,* ed. Barbara Bender, 1–17. Providence, RI: Berg.

———. 1993b. *Landscape: Politics and perspectives.* Providence, RI: Berg.

———. 2002a. Contested landscapes: Medieval to present day. In *The Material Culture Reader,* ed. Victor Buchli, 141–174. New York: Berg.

———. 2002b. Landscape and politics. In *The Material Culture Reader,* ed. Victor Buchli, 135–140. New York: Berg.

Bender, Barbara, and Margot Winer. 2001. *Contested landscapes: Movement, exile and place.* Oxford: Berg.

Bergman, Sten. 1937. *I morgonstillhetens land: Skildringar från en forskningsfärd till Korea* (In the land of morning calm: Descriptions from a research trip to Korea). Stockholm: A. Bonnier.

Bishop, Isabella Bird. [1898] 1970. *Korea and her neighbours.* Seoul: Yonsei University Press.

Bray, Francesca. 1994. Agriculture for developing nations. *Scientific American* 271(1): 30–37.

Breckenridge, Carol. 1989. The aesthetics and politics of colonial collecting: India at world fairs. *Comparative Studies in Society and History* 31(2): 195–216.

Brooks, Barbara. 2005. Reading the Japanese colonial archive: Gender and bourgeois civility in Korea and Manchuria before 1932. Chapter 8 in *Gendering modern Japanese history,* ed. Barbara Molony and Kathleen Uno. Cambridge: Harvard University Press.

Burke, Edmund. [1757] 1958. *A philosophical enquiry into the origin of our ideas of the sublime and beautiful.* London: Routledge and Kegan Paul.

Buswell, Robert. 1997. Monastery lay associations in contemporary Korean Buddhism: A study of the Puril Hoe. In *Religion and society in contemporary Korea,* ed. Lewis R. Lancaster, Richard Karl Payne, and Karen M. Andrews, 101–126. Korea Research Monograph 24. Berkeley: Institute of East Asian Studies, University of California.

———. 1999. Buddhism under Confucian domination: The synthetic vision of Sŏsan Hyujŏng. In *Culture and state in the late Chosŏn Korea,* ed. JaHyun Kim Haboush and Martina Deuchler, 134–159. Cambridge: Harvard University Press.

Butler, Judith. 1997. Against proper objects. In *Feminism meets queer theory,* ed. Elizabeth Weed and Naomi Schor, 1–30. Bloomington: Indiana University Press.

———. 1999. *Gender trouble: Feminism and the subversion of identity.* New York: Routledge. Originally published in 1990.

Callon, Michel, and John Law. 2004. Introduction: Absence-presence, circulation, and encountering in complex space. *Environment and Planning D: Society and Space* 22: 3–11.

Castles, Stephen. 2002. Migration and community formation under conditions of globalization. *International Migration Review* 36 (winter): 1,143–1,168.

Castles, Stephen, and Mark J. Miller. 1998. *The age of migration: International population movements in the modern world.* New York: Guilford Press.

Chandra, Vipan. 1988. *Imperialism, resistance, and reform in late nineteenth-century Korea: Enlightenment and the Independence Club.* Berkeley: Institute of East Asian Studies, University of California.

Chang Chu-kŭn. 1991. Song Sŏk-ha. In *Han'guk minjok munhwa taebaekkwa sajŏn* (Encyclopedia of Korean culture), Han'guk minjok munhwa taebaekkwa sajŏn p'yŏnch'anbu, 13: 18. Sŏngnam: Han'guk Chŏngsin Munhwa Yŏn'guwŏn.

Chapman, John. 1998. Objectification, embodiment and the value of places and things. In *The archaeology of value: Essays on prestige and the processes of valuation,* ed. Douglass W. Bailey and Steve Mills, 106–130. BAR International Series 730. Oxford: Archaeopress.

Cheju Provincial Bureau of Statistics. 1985. *Cheju t'onggye yŏnbo* (Cheju statistical yearbook). Cheju City: Cheju Provincial Bureau of Statistics.

Cheng, Sealing. 2002. Transnational desires: "Trafficked" Filipinas in U.S. military camp towns in South Korea. Ph.D. diss. School of Anthropology, University of Oxford.

Chesnutt, Michael. 1999. The many abodes of Olrik's epic laws. *Copenhagen Folklore Notes* 4: 7–11.

Ching, Leo T. S. 2001. *Becoming "Japanese": Colonial Taiwan and the politics of identity formation.* Berkeley: University of California Press.

Cho, Hae-joang. 1998. Constructing and deconstructing "Koreanness." In *Making majorities: Constituting the nation in Japan, Korea, China, Malaysia, Fiji, Turkey, and the United States,* ed. D. C. Gladney, 78–79. Stanford: Stanford University Press.

Cho, Uhn. 2004. Hybrid urbanity: Spatial embodiment and place politics. *Korea Journal* 44(3): 5–7.

Ch'oe Hyŏp. 1994. Honam munhwaron ŭi mosaek (In search of a theory of Chŏlla culture). *Han'guk Munhwa Illyuhak* (Korean cultural anthropology) 25: 29–45.

Ch'oe In-hak. 1995. *Minsokhak ŭi ihae* (Understanding folklore). Seoul: Miral.

Ch'oe Kil-sŏng. 1974. Misin t'ap'ae taehan ilgoch'al (A study of the destruction of superstition). *Han'guk minsokhak* (Korean folklore) 7: 39–54.

Ch'oe Yun. 1994. Hanak'o nŭn ŏpta: Sokssagim, sokssagim (There is no Hanak'o: Whisper, whisper). In *Yi Sang munhaksang chakp'umjip* (Collection of Yi Sang literary award literature), Yi Sang munhaksang sŏn'go wiwŏnhoe, 11–42. Seoul: Munhak sasangsa.

Chohon'gakchi. 2004. (The relic site of Chohon'gak). Seoul: Sadanbŏpin Sungmohoe.

Ch'oi Chang-jo. 1997. *Han'guk ŭi chaseang p'ungsu* (Indigenous p'ungsu in Korea). Seoul: Minumsa.

Choi, Chungmoo. 1993a. Transnational capitalism, national imaginary and the protest theater in South Korea. *Boundary* 2: 235–261.

———. 1993b. The discourse of decolonization and popular memory: South Korea. *Positions* 1: 77–102.

———. 1995. The Minjung Culture movement and the construction of popular culture in Korea. In *South Korea's Minjung Movement: The culture and politics of dissidence,* ed. Kenneth Wells, 105–118. Honolulu: University of Hawai'i Press.

Chŏn Myŏng-an. 2001. K'ŏming aut (Coming out). *Gay munhak* 1: 9–73.

Chŏn U-yong. 1999a. Taehan cheguk-ilche ch'ogi Sŏnhyech'ŏng Ch'angnaejang ŭi hyŏngsŏng kwa chŏngae: Sŏul Namdaemun sijang ŭi sŏngnip kwŏngwi (The formation and development of the Sŏnhyech'ŏng warehouse during the Great Han Empire and early colonial period: Details on the establishment of Namdaemun Market). *Sŏulhak yŏngu* 12 (April): 63–99.

———. 1999b. Taehan chegukki-ilche ch'ogi Sŏul konggan ŭi pyŏnhwa wa kwŏllyŏk ŭi chihyang (Spatial changes and power dynamics in Seoul during the Great Han and early colonial periods). *Chŏnnong saron* 5: 39–72.

Chŏng Chae-jŏng. 1997. Tŏksugung chubyŏn kŭndaehwa ŭi chach'ŭi (Traces of the modernization of the area around Tŏksu Palace). In *Sŏul ŭi munhwa yusa t'ambanggi* (An inquiry into the cultural remains of Seoul). Seoul: Sŏulhak Yŏnguso.

Chong, Chinsok. 1980. A study on the *Maeil Sinbo* (Daily news): Public information policy of the Japanese imperialists and Korean journalism under Japanese imperialism. *Journal of Social Sciences and Humanities* 52 (December): 59–114.

Chŏng Chong-su. 1999. *P'ungsuroponŭn urimunhwa iyagi* (Our culture as seen through geomancy). Seoul: Ungjindatk'ŏm.

Chŏng Chong-su and Sŏ Hŏn-gang. 1996. *Kyeryongsan* (Kyeryong mountain). Seoul: Taewonsa.

Chŏng Kŭn-sik. 1996. Chiyŏk chŏngch'esŏng kwa sangjing chŏngch'i (Regional identity and symbolic politics). *Kyŏngje wa Sahoe* (Economy and society) 30: 140–166.

Chŏng Sang-u. 2001. 1910 nyŏndae ilche ŭi chibae nolli wa chisikinch'ŭng ŭi insik: "Ilsŏndongjoron" kwa "munmyŏnghwa" ŭl chungsim ŭro (The logic of Japanese colonial rule and the consciousness of the Korean intellectual class during the 1910s: Theories on "same ancestral roots of Japanese and Koreans" and "the road to civilization"). *Hanguk saron* 46 (December): 183–231.

Chŏng Yŏng-ho. 1967. Kyŏngju Namsan Puljŏk poyu (Preserving the Buddhist remains of Kyŏngju Namsan). *Sahakji* (Historical bulletin) 1: 101–124.

Chongnoguji. 1994. (A history of the district of Chongno). Seoul: Chongnogu.

Chosen Sotokufu. [1941] 1994. *Keishu Nanzan no busseki* (Buddhist relics of Kyŏngju Namsan). Seoul: Yŏlhwadang.

Chung, Byung-jo. 1997. The Buddhist lay movement in Korean society. In *Religion and society in contemporary Korea,* ed. Lewis R. Lancaster, Richard Karl Payne, and Karen M. Andrews, 87–99. Korea Research Monograph 24. Berkeley: Institute of East Asian Studies, University of California.

Chung, Chai-sik. 1988. Confucianism: Tradition and transformation in Korea. *Transactions of the Korea Branch of the Royal Asiatic Society* 63: 37–56.

Chung, Heesun, and Gregory Vreek. 1999. Research note: Pessimism and pragmatism: Agricultural trade liberalisation from the perspective of South Korean farmers. *Asia Pacific Viewpoint* 40(3): 271–284.

Clifford, Mark. 1998. *Troubled tiger: Businessmen, bureaucrats, and generals in South Korea.* Armonk, NY: M. E. Sharpe.

Cohen, Eric. 1979. A phenomenology of tourist experiences. *Sociology* 13: 179–201.

———. 1982. Thai girls and farang men: The edge of ambiguity. *Annals of Tourism Research* 9: 403–428.

———. 1986. Lovelorn farangs. *Anthropological Quarterly* 59(3): 115–127.

Covell, Alan Carter. 1986. *Folk art and magic: Shamanism in Korea.* Seoul: Hollym Corp.

Cumings, Bruce. 1992. Silent but deadly: Sexual subordination in the U.S.–Korean relationship. In *Let the good times roll: Prostitution and the U.S. military in Asia,* ed. Saundra Pollock Sturdevant and Brenda Stoltzfus, 169–175. New York: New Press.

———. 1997. *Korea's place in the sun: A modern history.* New York: W. W. Norton.

Daos Korea Co. 2004. Taosŭ Kich'a yŏk (Daos train station). From Web site www.daos.net/railway/local_main.htm.

de Certeau, Michel. 1984. *The practice of everyday life.* Trans. Steven Randall. Berkeley: University of California Press.

de Ceuster, Koen. 2000. The changing nature of national icons in the Seoul landscape. *Review of Korean Studies* 3(2): 73–103.

———. 2001. The nation exorcised: The historiography of collaboration in South Korea. *Korean Studies* 25(2): 207–242.

Delaney, Jill. 1992. Ritual space in the Canadian museum of civilisation: Consuming Canadian identity. In *Lifestyle shopping: The subject of consumption,* ed. Rob Shields, 136–148. London: Routledge.

Delissen, Alain. 2001. The aesthetic pasts of space (1960–1990). *Korean Studies* 25(2): 243–260.

Detienne, Marcel, and Jean Pierre Vernant. 1978. *Cunning intelligence in Greek culture and society.* Trans. Janet Lloyd. Chicago: University of Chicago Press.

Dilling, Marnie W. 1998. The marathoner and the turtle. In *Nationalism and the construction of Korean identity,* ed. Hyung-il Pai and Timothy R. Tangherlini, 166–197. Korea Research Monograph 25. Berkeley: Institute of East Asian Studies, University of California.

Dirlik, Arif. 1993. *What is in a rim? Critical perspectives on the Pacific Region idea.* Boulder, CO: Westview Press.

Duncan, James, and Nancy Duncan. 1988. (Re)reading the landscape. *Environment and Planning D: Society and Space* 6: 117–126.

Duncan, Nancy. 1986. Suburban landscapes and suburbanites: A structurationist perspective on residential land-use in northern Westchester County. Ph.D. diss. Department of Geography, Syracuse University.

Dundes, Alan. 1977. Who are the folk? In *Frontiers of folklore,* ed. William Russell Bascom, 17–35. Boulder, CO: Published by Westview Press for the American Association for the Advancement of Science, Washington, DC.

Durkheim, Emile. [1915] 1965. *The elementary forms of the religious life.* Trans. Joseph Ward Swain. New York: Free Press.

Eckert, Carter J. 1991. *Offspring of empire: The Koch'ang Kims and the colonial origins of Korean capitalism, 1876–1945.* Seattle: University of Washington Press.

Epstein, Stephen. 2000. Anarchy in the U.K., solidarity in the R.O.K.: Punk rock comes to Korea. *Acta Koreana* 3: 1–34.

Epstein, Stephen, and Timothy R. Tangherlini, directors. 2002. *Our nation: A Korean punk rock community.* Filmakers Library.

Evon, Gregory N. 2001. Contestations over Korean Buddhist identities: The "introduction" to the Kyŏngho-jip. *Review of Korean Studies* 4(1): 11–33.

Faist, Thomas. 2000. *The volume and dynamics of international migration and transnational social spaces.* Oxford: Clarendon Press.

Foucault, Michel. 1986. Of other spaces. *Diacritics* 16: 22–27.

———. 1990. *The history of sexuality,* vol. 1. New York: Vintage Books.

Friedman, Susan Stanford. 1998. *Mappings: Feminism and the cultural geographies of encounter.* Princeton, NJ: Princeton University Press.

Fuchigami, Tadasuke. 1911. Keijō Honmachi-dōri no dōro kakuchō ni tsuite (On widening Keijō's Honmachi Avenue). *Chōsen* 45 (September 1): 48–51.

Fujimori, Terunobu. 1990. *Meiji no Tokyo keikaku* (The planning of Meiji Tokyo). Tokyo: Iwanami Shoten.

Fujitani, Takashi. 1996. *Splendid monarchy: Power and pageantry in modern Japan.* Berkeley: University of California Press.

Geertz, Clifford. 1973. *The interpretation of cultures: Selected essays.* New York: Basic Books.

Gell, Alfred. 1996. Vogel's net: Traps as artwork and artworks as traps. *Journal of Material Culture* 1(1): 15–38.

———. 1998. *Art and agency: An anthropological theory.* Oxford: Clarendon Press.

Genkō teikoku keiji hōten. 1909. (Current imperial criminal law codes). Tokyo: Ganshōdō.

Gotō, Yasushi. 1998. Nihon tōchika no taihoku jōnai no gaiku keisei ni kansuru kenkyū" (A study about the block formation inside Tāi-pak-siā under Japanese rule). *Dobokushi kenkyū* (Historical studies in civil engineering) 18: 103–116.

Graburn, Nelson. 1995. Tourism, modernity and nostalgia. In *The future of anthropology: Its relevance to the contemporary world,* ed. Akbar S. Ahmed and Cris N. Shore, 158–178. London: Athlone.

Greenough, Paul R., and Anna Lowenhaupt Tsing, ed. 2003. *Nature in the global south: Environmental projects in South and Southeast Asia.* Durham, NC: Duke University Press.

Gregerson, Edgar A. 1996. Sexual orientation. In *Encyclopedia of cultural anthropology,* vol. 4, ed. David Levinson and Melvin Ember, 1,173–1,178. New York: Henry Holt and Company.

Grindstaff, Beverly K. 1999. Creating identity: Exhibiting the Philippines at the 1904 Louisiana Purchase Exposition. *National Identities* 1(3): 245–264.

Grove, Richard. 1995. *Green imperialism: Colonial expansion, tropical island Edens, and the origins of environmentalism, 1600–1860.* Cambridge: Cambridge University Press.

Ha Yo-gil. 1980. *The National Folklore Museum.* Seoul: T'ongch'on Munhwasa.

Hacking, Ian. 1999. *The social construction of what?* Cambridge: Harvard University Press.

———. 2004. *Historical ontology.* Cambridge: Harvard University Press.

Halbwachs, Maurice. 1992. On collective memory. In *On collective memory,* ed. Maurice Halbwachs and Lewis A. Coser, 193–235. Chicago: University of Chicago Press.

Han Wu-gŭn. 1994. Tonghaksa-Tonghaksŏwŏn ŭi "Tonghak" sŏgŭisigo (The interpretation of "Eastern Learning" in Tonghak Temple and Tonghak Academy). *Minjok munhwa* 16: 3–22.

Han Yang-myŏng. 1996. Sŏknam Song Sŏk-ha minsok yŏn'gu wa minsokhak sajŏk wisang (About the folklore research of Song Sŏk-ha and the preservation of folklore). *Han'guk minsokhak* (Korean folklore) 28: 65–83.

Han'guk minsokch'on. 1980. (The Korean folk village). Seoul: Yangso.

———. 1997. (The Korean folk village). Seoul: K'otik'ŏm.

Harootunian, Harry D. 1974. Between politics and culture: Authority and the intellectual choice in Imperial Japan. In *Japan in crisis: Essays on Taishō democracy,* ed. Bernard S. Silberman, Harry D. Harootunian, and Gail Lee Bernstein, 110–155. Princeton, NJ: Princeton University Press.

Harvey, David. 1989. *The condition of postmodernity: An enquiry into the origins of cultural change.* Oxford: Blackwell.

———. 1993. From space to place and back again: Reflections on the condition of postmodernity. In *Mapping the futures: Local cultures, global change,* ed. Jon Bird, 3–29. London: Routledge.

Harvey, Youngsook Kim. 1979. *Six Korean women: The socialization of shamans.* St. Paul, MN: West Publishing.

Heidegger, Martin. 1993. Building dwelling thinking. In *Basic writings,* revised and expanded edition, ed. David Farrell Krell, 343–363. San Francisco: Harper San Francisco.

Hendry, Joy. 2000. *The Orient strikes back: A global view of cultural display*. Oxford: Berg.

Henry, Todd. 2004. Respatializing Chosŏn's royal capital: The politics of urban reforms and city life in early colonial Seoul. Conference presentation; unpublished manuscript.

———. 2005. Sanitizing empire: Japanese articulations of Korean otherness and the construction of early colonial Seoul, 1905–19. *Journal of Asian Studies* 64(3): 639–676.

Herzfeld, Michael. 1991. *A place in history: Social and monumental time in a Cretan town*. Princeton, NJ: Princeton University Press.

Hirsch, Eric. 1995. Landscape: Between place and space. In *The anthropology of landscape: Perspectives on place and space*, ed. Eric Hirsch and Michael O'Hanlon, 1–30. Oxford: Clarendon Press, Oxford University Press.

Hirsch, Eric, and Michael O'Hanlon, ed. 1995. *The anthropology of landscape: Perspectives on place and space*. Oxford: Clarendon Press, Oxford University Press.

Hitchcock, Michael. 1995. Indonesian cultural village museum and its forbears. *Journal of Museum Ethnography* 7: 17–24.

Hŏ Yŏng-sŏp. 1996. *Chosŏn Ch'ongdokpu, ku ch'ŏngsa kŏllip ŭi iyagi* (The construction of the former Japanese Government-General Building). Seoul: Hanul.

Hochschild, Arlie Russell. 1983. *The managed heart: Commercialization of human feeling*. Berkeley: University of California Press.

Hong Sun-min. 2004. Iljeŏŭi sikminch'imt'alkwa Kyŏngbokgung hweson: T'ongchikwolyŏkŭi saljingsŏng T'alchi (A study on Japan's colonial policy and destruction of Gyeongbokgung Palace: Destruction of the symbol of the sovereignty). *Munmyŏungyŏnji* 5(1): 5–34.

hooks, bell. 1990. Homeplace: A site of resistance. In *Yearning: Race, gender and cultural politics*, ed. bell hooks, 41–50. Boston: South End Press.

Huang Ru-si. 1988. *Han'gukin ŭi kut kwa mudang* (Koreans' shamans and shaman rituals). Seoul: Munhwasa.

Humphrey, Caroline. 1995. Chiefly and shamanist landscapes in Mongolia. In *The anthropology of landscape: Perspectives on place and space*, ed. Eric Hirsch and Michael O'Hanlon, 135–162. Oxford: Clarendon Press, Oxford University Press.

Hung, Wu. 1991. Tiananmen Square: A political history of monuments. *Representations* 35: 84–117.

Hwang, Kyung Moon. 2004. *Beyond birth: Social status in the emergence of modern Korea*. Cambridge: Harvard University Asia Center. Dist. Harvard University Press.

Hwang Su-yŏng. 1994. Haeje (Introduction). In *Keishu Nanzan no busseki* (Buddhist relics of Kyŏngju Namsan), ed. Chosen Sotokufu, unpaginated introductory section. Seoul: Yŏlhwadang.

Ika ni seba shumika shiuruka. 1915. (How to improve the taste [of Seoul]). *Chōsen oyobi manshū* (March 1).

Im Chae-hae. 1991. *Han'guk minsok kwa chŏnt'ong ŭi segye* (The realm of Korean folklore and tradition). Seoul: Chisik Sanopsa.

Imperial Rescript. 1910. Heigō no shosho (Imperial Rescript on education). August 29.

In Kwŏn-hwan. 1978. *Han'guk minsok haksa* (History of Korean folklore). Seoul: Yŏrhwadang.

Ishida, Osamu. 1998. "Dōka" seisaku to tsukurareta kannen toshite no "Nihon" ("Assimilation" policy and the constructed concept of "Japan"). *Shisō* 892 (October): 47–75.

Ishida, Yorifusa. 1987. *Nihon kindai toshi keikakushi kenkyū* (Studies on the history of Japanese urban planning). Tokyo: Kashiwa shobō.

Jacobs, Jane M. 1996. *Edge of empire: Postcolonialism and the city.* London: Routledge.

Jager, Sheila Miyoshi. 2003. *Narratives of nation building in Korea: A genealogy of patriotism.* Armonk, NY: M. E. Sharpe.

Jagose, Annamarie. 1996. *Queer theory: An introduction.* New York: New York University Press.

Janelli, Roger. 1988. The origins of Korean folklore scholarship. *Journal of American Folklore* 99: 24–49.

Janelli, Roger L., with Dawnhee Yim. 1993. *Making capitalism: The social and cultural construction of a South Korean conglomerate.* Stanford, CA: Stanford University Press.

Jones, George Heber. 1907. *Korea: The land, people, and customs.* Cincinnati: Jennings and Graham.

Ju Yun-jŏng. 2003. Chosŏn mulsan kongjinhoe wa sikminjuŭi sisŏn (Chosŏn mulsan kongjinhoe and colonial gaze). *Munhwa kwahak* 33: 145–160.

Jung Hye-Seung. 1996. KBS News 9. *KBS News,* September 9.

Jung Ki-yong. 1992. From Kwanghwa Gate to Sejong Street. *Munhwa kwahak* (Culture science) 5: 45–64.

Kang In-jun. 1990. Namsan e orŭmyŏnsŏ (While climbing Namsan). *Puch'ŏnim Maŭl* (Buddha Village) 55: 13.

Kang In-sŏn. 1993. Chosŏn Ch'ongdokbu ch'ŏngsa ch'ŏlgŏnonjaengŭn uri ege muŏtŏl namgyŏtna (What is the consequence of the debates surrounding the destruction of the former Government-General Building?). *Wolgan Chosŏn* 163: 334–344.

Kang Myŏng-gwan. 1992. Chosŏn hugi Soŭl ŭi chunggan kyech'ŭng kwa yuhŭng paltal (The development of the middling classes and the development of entertainment during the late Chosŏn dynasty). *Minjok munhak yŏngu* (Studies in national literature) 2, 181–207.

Kang Nae-hŭi. 2002. Hankukŭi sikminji kŭndaesŏngkwa chungkyŏkŭi pŏnyŏk (Colonial modernity in Korea and the translation of a shock). *Munhwa kwahak* 31: 74–97.

Kang Pyŏng-sik. 1994. *Ilche sidae Sŏulŭi t'oji yŏngu* (Study on the land in Seoul during Japanese colonial rule). Seoul: Minjok Munhwasa.

Kang Sin-yong. 2004. Kankoku ni okeru kindai toshi kōen no juyō to tenkai (The adoption and development of modern city parks in Korea). Ph.D. diss. Kyoto University.

Kant, Immanuel. [1790] 1952. *The critique of judgment.* Trans. James Creed Meredith. Oxford: Clarendon Press.

Kaplan, Flora Edouwaye S., ed. 1994. *Museums and the making of "Ourselves": The role of objects in national identity.* London: Leicester University Press.

Kasbarian, John A. 1996. Mapping Edward Said: Geography, identity, and the politics of location. *Environment and Planning D: Society and Space* 14(5): 529–557.

Keim, Willard D. 1979. *The Korean peasant at the crossroads: A study in attitudes.* Bellingham: Western Washington University.

Keimu ihō. 1911. (Police digest). Keijō.

———. 1913. (Police digest). Keijō.

Keith, Michael. 1997. Conclusion: A changing space and a time for change. In *Geographies of resistance,* ed. Steve Pile and Michael Keith, 277–286. London: Routledge.

Kellerman, Aharon, and Maoz Azaryahu. 1994. Symbolic places of national history: A study of Zionist mythical geography. *Transactions of Institute of British Geographers* 24: 109–123.

Kendall, Laurel. 1977. Receiving the *samsin* grandmother. *Transactions of the Korea Branch of the Royal Asiatic Society* 52: 55–70.

———. 1985. *Shamans, housewives, and other restless spirits: Women in Korean ritual life.* Honolulu: University of Hawai'i Press.

———. 1988. *The life and hard times of a Korean shaman: Of tales and the telling of tales.* Honolulu: University of Hawai'i Press.

———. 1996. Korean shamans and the spirits of capitalism. *American Anthropologist* 98: 512–527.

———. 1998. The shaman's journey: Real and ideal in a living folk tradition. In *Korean shamanism: Revivals, survivals, and change,* ed. Keith Howard, 91–111. Seoul: Royal Asiatic Society, Korea Branch.

———. 1999. Peoples under glass: A tale of two museums. In *Consuming ethnicity and nationalism: Asian experiences,* ed. Kosaku Yoshino, 111–132. Honolulu: University of Hawai'i Press.

———. 2002. An old shaman in a tiled roof house. *Shaman* 10(1-2): 113–124.

Kim Chin-bong. 2000. *Samil Undongsa yŏn'gu* (Studies on the history of the March First Movement). Seoul: Kukhak Charyowŏn.

Kim Chŏng-dong. 2000. *Nama innŭn yŏksa, sarajinŭn kŏnch'ungmul* (History that remains, buildings that vanish). Seoul: Taewŏnsa.

———. 2004. *Kojong hwangje ka sarang han Chŏngdong kwa Tŏksugung* (The neighborhood of Chŏngdong and Tŏksu Palace which Emperor Kojong loved). Seoul: Palŏn.

Kim Dong-hun. 2002. The dream of Lee Kang-suk. *Han'gyŏreh sinmun,* May 23.

Kim, Eun Mee. 1997. *Big business, strong state: Collusion and conflict in South Korean development, 1960–1990.* Albany: State University of New York Press.

Kim, Hong-nam. 1998. Removing the legacy of the Korean past. *Curator* 41(3): 178–186.

Kim, Hyun-sook. 1998. "Yanggongju" as an allegory of the nation: The representation of working-class women in popular and radical texts. In *Dangerous women: Gender and Korean nationalism,* ed. Elaine H. Kim and Chungmoo Choi, 175–202. New York: Routledge.

Kim In-sik, writer and director; Kim Jae-won, producer. 2002. *Rodŭ mubi* (Road movie). Seoul: Ent'o Won. DVD.

Kim Ki-bin. 1993. *600 nyŏn Sŏul Ttangirŭm Iyagi* (600 years of Seoul place names). Seoul: Sallimt'o.

Kim Ki-ho. 1995. Ilche sidae ch'ogi ŭi tosi kyehoek e taehan yŏngu: Kyŏngsŏngpu sigu kaejŏng ŭl chungsim ŭro (A study on the city planning of Seoul during the early colonial period with a focus on urban reforms). *Sŏulhak yŏngu* 6: 41–66.

Kim Kwang-ŏk. 1994a. Munhwa kongdongch'e wa chibang chŏngch'i: Ssijok ŭi kujo rŭl chungsim ŭro (Cultural community and local politics: Centering on the structure of lineages). *Han'guk munhwa illyuhak* (Korean cultural anthropology) 25: 116–182.

———. 1994b. Rituals of resistance: The manipulation of shamanism in contemporary Korea. In *Asian visions of authority: Religion and the modern states of East and Southeast Asia,* ed. Charles F. Keyes, Laurel Kendall, and Helen Hardacre, 195–221. Honolulu: University of Hawai'i Press.

Kim Kwang-u. 1990. Taehan cheguk sidae ŭi tosi kyehoek (Urban planning during the Great Han Empire period). *Hyangt'o Sŏul* 50, 95–122.

Kim, Pu-sik. [12th century] 1998. *Samguk sagi.* Ed. Yi Kang-nae. Original written in the twelfth century CE. Seoul: Han'gilsa.

Kim Sang-hyŏn. 1995. Yŏksa tosi Kyŏngju ŭi munhwa yusallon (An essay on the cultural remains of the historic city of Kyŏngju). In *Han'guk Pulgyosa sanch'aek* (A walk through Korean Buddhist history), 278–309. Seoul: Uri Ch'ulp'ansa.

Kim Seong-nae (Sŏng-rye) 2000. P'ungsuwa sikminjuŭi kiŏgŭi erotic chŏngch'ihak (P'ungsu and the erotic politics of colonial memory). *Han'guk chongkyoyŏn'gu* (Research on Korean religion) 2: 123–157.

Kim, Seung Kyung, and John Finch. 2002. Living with rhetoric, living against rhetoric: Korean families and the IMF economic crisis. *Korean Studies* 26(1): 120–139.

Kim Sŏn-p'ung. 1991. Minsokhak (Folklore). In *Han'guk minjok munhwa taebaekkwa sajŏn* (Encyclopedia of Korean culture), Han'guk minjok munhwa taebaekkwa sajŏn p'yŏnch'anbu, 8: 750–751. Sŏngnam: Han'guk Chŏngsin Munhwa Yŏn'guwŏn.

Kim Sŏn-p'ung, ed. 1994. *Han'guk minsokhak ŭi ihae* (Understanding Korean folklore). Seoul: Munhak Ak'ademi.

Kim Tae-kyong, ed. 1988. *Korean people's block prints: Selected representative works.* Seoul: Hong Hyong-Suk.

Kim Tae-sŭng, director. 2001. *Beonjijeompeureul hada* (Bungee jumping of their own). Seoul: Eye Entertainment.

Kim Tŏk-su. 1983. Saudi SAFWA esŏ Kim Tŏk-su p'yŏnji (Kim Tŏk-su's letter from SAFWA, Saudi Arabia). *Pŏppo* (Dharma news) 162: 3.

———. 1984. Puch'ŏnim kke kamsa tŭrimyŏnsŏ . . . (While giving thanks to Buddha . . .). *Puch'ŏnim Maŭl* (Buddha Village) 1: 1–2.

———. 1986. Kyŏngju Namsan Puljŏk sullye I (Pilgrimage to the Buddhist relics of Kyŏngju Namsan, Part I). *Puch'ŏnim Maŭl* (Buddha Village) 20: 2–8.

Kim, Tong-ni. 1971. Portrait of a shaman. In *Flowers of fire: Twentieth-century Korean stories,* ed. and trans. Peter H. Lee, 58–90. Honolulu: University of Hawai'i Press.

Kim Tong-uk, Ch'oe In-hak, Ch'oe Kil-sŏng, and Ch'oe Rae-ok. 1988. *Han'guk minsokhak* (Korean folklore). Seoul: Saemunsa.

Kim Yong-ok. 2003. *To'ol ŭi Ch'ŏnggyech'ŏn iyagi* (To'ol on the subject of the Ch'ŏnggye Stream). Seoul: T'ongnamu.

Kim Yong-sam. 1995. Is Kim Young-Sam government a P'ungsu government? *Wolgan Chosŏn* (Monthly Chosŏn), October: 156–179.

Kim Yong-suk. 1987. *Chosŏnjo kungjung p'ungsok yŏn'gu* (A study of palace customs in the Chosŏn dynasty). Seoul: Iljisa.

Kim Yoo-hyuk, ed. 1981. *Saemaul undong: Determination and capability of the Koreans.* Trans. H. Han et al. Seoul: Saemaŭl Yŏnguhoe (Institute of Saemaul Studies).

Kim Young-kwon and Lee Soo-jung, ed. 1979. *Handbook of Korea.* Seoul: Korean Overseas Information Service.

Kim, Young-na. 2001. The first encounter with the West: Korean exhibit at World Columbian Exhibition, Chicago. Unpublished manuscript, Korea Institute, Harvard University.

Kim Yun-gŭn. 1995. Silla munhwa tonginhoe ŭi ŏje wa onŭl (The friends of Silla Culture, yesterday and today). *Kyŏngju munhwa* (Kyŏngju culture) 1: 276–283.

Kirk, Don. 2002. From Russia without "love": GI bars in South Korea. *International Herald Tribune,* September 20.

Ko Tong-hwan. 1997. The characteristics of the urban development of Seoul during the late Chosŏn dynasty with a focus on the changes in urban structure. *Seoul Journal of Korean Studies* 10 (December): 95–123.

———. 1999. 18 segi Sŏul ŭi tosi kujo pyŏnhwa wa ryuhang munhwa (Changes in the urban structure of eighteenth-century Seoul and commoner culture). *Sŏng'gok nonch'ong* 30: 327–372.

Korea Times. 1989. S.K. farmers clash with 4,000 police. *Korea Times (Los Angeles Edition),* February 14: 1.

———. 2002. Prosecution combats brothels near U.S. bases. *Korea Times,* October 11.

Korean mudang and pansu. 1903. *Korea Review* 3: 145–149, 203–208, 257–260, 301–305, 342–346, 383–389.

Korean Research Institute for Human Settlements. 2004. Korea's experience and KOICA's Program for Human Resource Development (HRD). From Web site http://www.krihs.re.kr/eng/cooperation/04_koica1.asp.

Ku Chung-hoe. 2001. *Kyeryongsan kuttang yŏn'gu* (A study of the shaman shrines on Kyeryong Mountain). Seoul: Kukhak Charyowŏn.

Kuchler, Susanne. 1993. Landscape as memory: The mapping of process and representation in Melanesian society. In *Landscape: Politics and perspectives,* ed. Barbara Bender, 85–106. Providence, RI: Berg.

Kungnip Minsok Pangmulgwan. 1986a. *Kungnip Minsok Pangmulgwan 50 nyŏnsa* (A 50-year history of the Korean National Folklore Museum). Seoul: Tongyang Munhwa Inswae.

———. 1986b. *Kungnip Minsok Pangmulgwan* (The National Folklore Museum). Seoul: Kungnip Minsok Pangmulgwan.

———. 1994. *Guide to the National Folk Museum, Seoul Korea.* Seoul: Shin-yoo.

———. 1996. *Kungnip Minsok Pangmulgwan* (The National Folklore Museum). Seoul: Kungnip Minsok Pangmulgwan.

Kuper, Hilda. 2003. The language of sites in the politics of space. In *The anthroplogy of space and place: Locating culture,* ed. S. M. Low and D. Lawrence-Zuniga, 247–263. Oxford: Blackwell.

Kwŏn Hyŏk-ch'ŏl. 1998. *Han'gyŏreh sinmun,* Local Section, September 22, 24.

Kwŏn Kam 1793. Kyeryongsan chugi (A postscript on Kyeryong Mountain). In *Chagi* (Self-writing): 5.

Kyeryongsi. 2004. *Kyeryongsi ŭi yŏnhyŏk kwa tosigyehoek* (The history and urban planning of Kyeryong City). Kyeryongsi: Kyeryongsi.

Kyŏngju Kaebal Kyehoektan. 1971. Kyŏngju kwan'gwang chonghap kaebal kyehoek (Kyŏngju tourism comprehensive development plan). Seoul: Ch'ŏngwadae. Unpublished manuscript.

Latour, Bruno. 1983. Give me a laboratory and I will raise the world. In *Science observed: Perspectives on the social study of science,* ed. Karin D. Knorr-Cetina and Michael Mulkay, 141–170. London: Sage.

———. 1996. On actor-network theory: A few clarifications. *Soziale Welt* 47: 369–381.

———. 1999. *Pandora's hope: Essays on the reality of science studies.* Cambridge: Harvard University Press.

Law, John. 1999. After ANT: Complexity, naming, and topology. In *Actor network theory and after,* ed. John Law and John Hassard, 1–14. Oxford: Blackwell.

———. 2002. *Aircraft stories: Decentering the object in technoscience.* Durham, NC: Duke University Press.

Lee Chong-min. 2002. 1910 nyŏndae Kyŏngsŏng chumindŭl ŭi 'choe' wa 'bŏl': Kyŏngbŏmjae t'ongje rŭl chungsim ŭro (Crime and punishment: On the control of minor offenses in 1910s Seoul). *Sŏulhak yŏngu* 15: 139–180.

———. 2004. Keihanzai no torishimari hōrei ni miru minshū tōsei: Chōsen no baai o chūsin ni (Control the populace as seen through the legal regulation of petty crimes: The case of Korea). In *Shokuminchi teikoku nihon no hōteki kōzō* (The legal structure of Imperial Japan in its colonies), ed. Toyomi Asano and Toshihiko Matsuda. Tokyo: Shinsansha.

Lee, Hye-kyung 1997. The employment of foreign workers in Korea: Issues and policy suggestions. *International Sociology* 12(3): 353–371.

Lee, Jeong-man. 1991. The generation of sacrifice: Modernization and Korean farmers. Ph.D. diss. (geography). University of California, Berkeley.

Lee Sang-hae. 1991. Kyŏngbokgung, Kyŏnghŭigung pokwŏnkwa yet Chosŏn Ch'ongdokbu ch'ŏngsa ch'ŏlgŏmunje (The issues of restoring the Kyŏngbok Palace and Kyŏnghŭi Palace and removing the old Government-General Building). *Kŏnch'uk* 35(2): 52–56.

Lefebvre, Henri. 1991. *The production of space.* Trans. Donald Nicholson-Smith. Oxford: Blackwell.

Lie, John. 1998. *Han unbound: The political economy of South Korea.* Stanford, CA: Stanford University Press.

Little, Paul E. 1999. Environments and environmentalisms in anthropological research: Facing a new millennium. *Annual Review of Anthropology* 28: 253–284.

MacCannell, Dean. 1976. *The tourist: A new theory of the leisure class.* New York: Schocken Books.

Maeil sinbo. 1910. Nodong ŭi sinsŏng (The sanctity of labor). *Maeil sinbo,* October 21.

———. 1910. Yasi ŭi p'ilyo (The need for night markets). *Maeil sinbo,* November 17.

———. 1911. Kyŏngsŏng kaksangjŏm (Keijō's merchant shops). *Maeil sinbo,* August 17.

———. 1911. Kyŏngsŏng puminhoe haesan (The dissolution of the Seoul people's association). *Maeil sinbo,* September 23.

———. 1911. Nanmunoe t'oji suyong (Land confiscation outside the South Gate). *Maeil sinbo,* October 25.

———. 1912. Kaok t'oji ŭi chŭngmyŏng (Certifying homes and property). *Maeil sinbo,* July 10.

———. 1912. Sigu kaejŏng (Urban reforms). *Maeil sinbo,* November 17.

———. 1912. Sigu kaejŏng kwa ohae (Misunderstandings of urban reforms). *Maeil sinbo,* November 17.

———. 1913. Sigaji kŏnch'uk ch'uich'e (Control of buildings in the city area). *Maeil sinbo,* February 28.

———. 1913. Sigu kaejŏng kwa konggongsim (Urban reforms and civic spirit). *Maeil sinbo,* August 24.

———. 1914. Kongwŏn e tae haya (On parks). *Maeil sinbo,* June 28.

———. 1917. Minp'ung kaesŏn (Reforming popular customs). *Maeil sinbo,* June 20–August 12.

Mason, David A. 1999a. *Spirit of the mountains: Korea's San-Shin and traditions of mountain-worship.* Elizabeth, NJ: Hollym.

———. 1999b. In memory of Zo Zayong, a great Korean. From Web site http://san-shin.org/Zo-01.html.

Matsui, Shigeru. 1913. Kankoku ni okeru shakai kyōiku to keisatsu tono kankei (Social education in Korea and its relationship with the police). In *Jichi to keisatsu* (Self-administration and the police). Tokyo: Keigansha.

Matsui, Yayori. 1995. The plight of Asian migrant women working in Japan's sex industry. In *Japanese women: New feminist perspectives on the past, present, and future,* ed. Kumiko Fujimura-Fanselow and Atsuko Kameda, 309–322. New York: Feminist Press.

McIntyre, Donald. 2002. Base instincts. *Time Asia Magazine* 160(5) (August 12).

McLelland, Mark. 2000. Male homosexuality and popular culture in modern Japan. *Intersections* 3, from Web site (accessed September 22, 2004) http://wwwsshe.murdoch.edu.au/intersections/issue3/mclelland2.html.

McMichael, William H. 2002. Sex slaves: How women are lured into South Korea's flesh trade. *Army Times/Navy Times (Military Times),* August 12.

Merrifield, Andy. 2006. *Henri Lefebvre: A critical introduction.* New York: Routledge.

Miller, Daniel. 2000. The fame of Trinis: Websites as traps. *Journal of Material Culture* 5(1): 5–24.

Ministry of Home Affairs. 1980. *Ten year history of the Saemaul Undong.* Seoul: Ministry of Home Affairs.

Ministry of Transportation. 1994. *Korean tourism annual report, 1993.* Seoul: Government Printing Office.

Mitchell, Timothy. 1989. The world as exhibition. *Comparative Studies in Society and History* 31: 217–236.

Mizuno, Naoki. 2002. Chōsen shokuminchi shihai to namae no "saika": "Naichijin ni seimiei" no kinshi o megutte (Colonial rule of Korea and name "differentiation": On the prohibition of full names resembling those of Japanese). In *Shokuminchi shugi to jinruigaku* (Colonialism and anthropology), ed. Katsuhiko Yamaji and Masakazu Tanaka, vi, 567. Nishinomiya: Kansai daigaku shuppankai.

Mok Su-hyŏn. 2000. Ilcheha pangmulgwan ŭi hyŏngsŏng kwa kŭ ŭimi (The establishment of museums under Japanese colonial rule and their meaning). Ph.D. diss. Seoul National University.

Montgomery, Heather. 2001. *Modern Babylon? Prostituting children in Thailand.* New York: Berghahn Books.

Moon, Katharine H. S. 1997. *Sex among allies: Military prostitution in U.S.–Korea relations.* New York: Columbia University Press.

———. 2000. Strangers in the midst of globalization: Migrant workers and Korean nationalism. In *Korea's globalization,* ed. Samuel S. Kim, 147–169. Cambridge Asia-Pacific Studies. New York: Cambridge University Press.

Moon, Okpyo. 2001. Tourist re-making of a Confucian elite village in Korea: The case of Hahoe in Andong. *Korean Social Science Journal* 28: 1–28.

Morphy, Howard. 1993. Colonialism, history and the construction of place: The politics of landscape in northern Australia. In *Landscape: Politics and perspectives,* ed. Barbara Bender, 205–243. Providence, RI: Berg.

Munhwajaekwallikuk Statistical Overview. 1997; updated 2004.

Munn, Nancy. 2003. Excluded spaces: The figure in the Australian aboriginal landscape. In *The anthropology of space and place: Locating culture,* ed. S. M. Low and D. Lawrence-Zuniga, 92–109. Oxford: Blackwell.

Nam Ha-jŏng 1731. Kyeryong kihaeng (A travelogue to Kyeryong Mountain). Poem in *Dongsoyugo,* vol. 4.

Needham, Joseph, and Ling Wang. 1965. *Science and civilisation in China.* Cambridge: Cambridge University Press.

Nemeth, David J. 1987. *The architecture of ideology: Neo-Confucian imprinting on Cheju Island, Korea.* Berkeley: University of California Press.

———. 1988. The walking tractor: Trojan horse in the Cheju Island landscape. *Korean Studies* 12: 14–38.

Niessen, Sandra A., Ann Marie Leshkowich, and Carla Jones, ed. 2003. *Re-orienting fashion: The globalization of Asian dress.* Oxford: Berg.

Ning, Wang. 1999. Rethinking authenticity in tourism experience. *Annals of Tourism Research* 25: 349–370.

No In-hwa. 1980. Taehan cheguk sigi ŭi Hansŏng chŏngi hoesa e taehan yŏngu: Kwangmu kaehyŏk kwa migukch'ŭk igwŏn ŭi ilyangt'ae (A study of the Seoul Electric Company during the Great Han Empire period: The Kwangmu reforms and one aspect of American concessions). *Idae sawŏn* 17: 1–27.

Nora, Pierre. 1989. Between memory and history: Les Lieux de Mémoire. *Representations* 26: 7–25.

Ŏ Suk-kwŏn. 1971. P'aegwan chapki (A storyteller's miscellany). In *Kugyŏk taedong yasŭng* (Unofficial history of the great eastern country, in translation), ed. Minjok Munhwa Ch'ujinhoe, 417–554 and 728–784. Seoul: Minjok Munhwa Ch'ujinhoe.

Oguma, Eiji. 1995. *Tan'itsu minzoku shinwa no kigen: "Nihonjin" no jigazōno keifu* (The origins of the myth of ethnic homogeneity: The genealogy of "Japanese" self-images). Tokyo: Shin'yōsha.

Oh Myŏng-sŏk. 1998. 1960–70 nyŏndae ŭi munhwachŏngch'aekwa minjok munhwa tamnon (Cultural policy and the discourse of national culture in 1960–70). *Pikyomunhwayŏn'gu* 4: 121–151.

Okita, Kinjō. 1905. *Rimen no kankoku* (Korea behind the mask). Osaka: Kibunkan.

Onishi, Norimitsu. 2004. This 21st-century Japan, more contented than driven. *New York Times,* February 4: A4.

Osborne, Brian S. 1998. Constructing landscapes of power: The George Etienne Cartier monument, Montreal. *Journal of Historical Geography* 24(4): 431–458.

Ōta, Hideharu. 2003. Kŭndae hanil yangguk ŭi sŏnggwak insik kwa Ilbon ŭi Chosŏn singminji chŏngch'aek (Modern Korean and Japanese perceptions of castle walls and Japan's colonial policy in Korea). *Hanguk saron* 49 (June): 185–230.

Pai, Hyung Il. 2000. *Constructing "Korean" origins: A critical review of archaeology, historiography, and racial myth in Korean state-formation theories.* Cambridge: Harvard University Asia Center.

———. 2001. The creation of national treasures and monuments: The 1916 Japanese laws on the preservation of Korean remains and relics and their colonial legacies. *Journal of Korean Studies* 25: 72–95.

Pak Chong-in. 1996. Orŭl ttae mada saeroun kkaedarŭm (A new enlightenment every time one goes up). *Puch'ŏnim Maul* (Buddha Village) 110: 2.

Pak Hŭng-ju. 2001. *Sŏul ŭi maŭl kut* (Seoul's village rituals). Seoul: Somundang.

Pak, Yangjin. 1997. Korea's national museum and colonial experience. *Cultural Survival Quarterly* 21: 30–32.

Park Kŭn-ae. 1998. Nosukcha chahwal topki ponkyŏkhwa (Actualization of promoting homeless self-support). *Han'gyŏreh sinmun,* Local Section, September 18: 20.

Park Yong-hyŏn. 1999. Ai-em-epŭ ka tturŏ non chukŭm ŭi tŏnŏl (Tunnel of the death made by the IMF crisis). *Han'gyŏreh sinmun* 21(249), March 18. Web site http://www.hani.co.kr/h21/data/L990308/1paq3802.html.

Peattie, Mark R. 1984. Japanese attitudes toward colonialism, 1895–1945. In *The Japanese colonial empire, 1895–1945,* ed. Ramon Hawley Myers, Mark R. Peattie, and Ching-chih Chen, 80–128. Princeton, NJ: Princeton University Press.

Pettid, Michael J. 2001. Sexual identity in Chosŏn period literature: Humorous accounts of forbidden passion. *Review of Korean Studies* 4(1): 61–86.

Pile, Steve. 1997. Introduction: Opposition, political identities and spaces of resistance. In *Geographies of resistance,* ed. Steve Pile and Michael Keith, 1–32. London: Routledge.

Pile, Steve, and Michael Keith, ed. 1997. *Geographies of resistance.* London: Routledge.

Pile, Steve, and N. J. Thrift. 1995. *Mapping the subject: Geographies of cultural transformation.* London: Routledge.

Pitts, Forrest R. 2002. Sliding sideways into geography. In *Geographical voices: Fourteen autobiographical essays,* ed. Peter Gould and Forrest R. Pitts, 269–292. Syracuse, NY: Syracuse University Press.

———. 2004. Personal interview, July 30. Unpublished manuscript.

Pratt, Keith L., and Richard Rutt, with additional material by James Hoare. 1999. *Korea: A historical and cultural dictionary.* Richmond, UK: Curzon Press.

Puch'ŏnim Maŭl. 1987. Saero naon ch'aek (New books). *Puch'ŏnim Maŭl* (Buddha Village) 29: 11.

Purdum, Todd S. 2003. Walt Rostow, advisor to Kennedy and Johnson, dies at 86. *New York Times,* February 15: A23.

Pyo Min-su, director. 1999. *Sŭlpŭn yuhok* (Sad temptation). KBS Drama Production Studio.

Raffles, Hugh. 2002a. *In Amazonia: A natural history.* Princeton, NJ: Princeton University Press.

———. 2002b. Intimate knowledge. *International Social Science Journal* 54(173): 325–334.

Richter, Linda K. 1999. The politics of heritage tourism: Emerging issues for the new millennium. In *Contemporary issues in tourism development,* ed. Douglas G. Pearce and Richard W. Butler, 108–126. London: Routledge.

Robinson, Michael E. 1986. Nationalism and the Korean tradition, 1896–1920: Iconoclasm, reform, and national identity. *Korean Studies* 10: 35–53.

———. 1988. *Cultural nationalism in colonial Korea, 1920–1925.* Seattle: University of Washington Press.

Rostow, Walt W. [1961] 1990. *The stages of economic growth: A non-communist manifesto.* Cambridge: Cambridge University Press.

———. 2003. *Concept and controversy: Sixty years of taking ideas to market.* Austin: University of Texas Press.

Routledge, Paul. 1993. *Terrains of resistance: Nonviolent social movements and the contestation of place in India.* Westport, CT: Praeger.

———. 1997. The imagineering of resistance: Pollok Free State and the practice of postmodern politics. *Transactions of the Institute of British Geographers,* New Series 22: 359–376.

Ruhlen, Rebecca N. 2003. Korean alterations: Nationalism, social consciousness, and "traditional" clothing. In *Re-orienting fashion: The globalization of Asian dress,* ed. S. A. Niessen, Ann Marie Leshkowich, and Carla Jones, 117–137. Oxford: Berg.

Ryan, Chris, and Colin Michael Hall. 2001. *Sex tourism: Marginal people and liminalities.* London: Routledge.

Ryu, Je-Hun. 2000. *Reading the Korean cultural landscape.* Elizabeth, NJ: Hollym.

Said, Edward W. 1978/1979. *Orientalism.* New York: Vintage Books.

———. 1993. *Culture and imperialism.* New York: Knopf.

Sandberg, Mark. 1995. Effigy and narrative: Looking into the nineteenth-century folk museum. In *Cinema and the invention of modern life,* ed. Leo Charney and Vanessa R. Schwartz, 320–361. Berkeley: University of California Press.

———. 2003. *Living pictures, missing persons: Mannequins, museums, and modernity.* Princeton, NJ: Princeton University Press.

Santos, Aida F. 1992. Gathering the dust: The bases issue in the Philippines. In *Let the good times roll: Prostitution and the U.S. military in Asia,* ed. Saundra Pollock Sturdevant and Brenda Stolzfus, 32–44. New York: Free Press.

Schmid, Andre. 2002. *Korea between empires, 1895–1919.* New York: Columbia University Press.

Schramm, Wilbur Lang, and Daniel Lerner. 1976. *Communication and change: The last ten years—and the next.* Honolulu: University of Hawai'i Press.

Scott, David. 1995. Colonial governmentality. *Social Text* 43: 191–220.

Scott, James C. 1985. *Weapons of the weak: Everyday forms of peasant resistance.* New Haven: Yale University Press.

———. 1998. *Seeing like a state: How certain schemes to improve the human condition have failed.* New Haven: Yale University Press.

Sejong sillok. 1968–. *Sejong sillok (The annals of King Sejong).* Seoul: Sejong Taewang Kinyŏm Saŏphoe. (Volume and page numbers as cited in text.)

Seo Hyŏn-ju. 2002. Ilchemal ilcheha Sŏul ŭi habu haengjŏng chedo yŏngu: Chŏngdonghoe wa ch'ongdae rŭl chungsim ŭro (A study on the lower administrative system in the late Chosŏn and Japanese colonial periods: Neighborhood associations and government-appointed administrators). Ph.D. diss. Seoul National University.

Sharp, Joanne. 2000. *Entanglements of power: Geographies of domination/resistance.* London: Routledge.

Shields, Rob. 1991. *Places on the margin: Alternative geographies of modernity.* London: Routledge Chapman Hall.

Shim U-sŏng. 1994. Namsadang. In *Han'guk minjok munhwa taebaekkwa sajŏn,* 5: 408–411. Sŏngnam: Han'guk Chŏngsin Munhwa Yŏn'guwŏn.

Shin, Gi-wook. 1998. Nation, history, and politics: Rhetoric of minjok in postwar Korea. In *Nationalism and the construction of Korean identity,* ed. Hyung Il Pai and Timothy R. Tangherlini, 148–165. Korea Research Monograph 25. Berkeley: Institute of East Asian Studies, University of California.

Shin, Gi-wook, James Freda, and Gihong Yi. 1999. The politics of ethnic nationalism in divided Korea. *Nations and Nationalism* 5(4): 465–484.

Shin, Gi-wook, and Michael E. Robinson. 1999a. Introduction: Rethinking colonial Korea. In *Colonial modernity in Korea,* ed. Gi-wook Shin and Michael Robinson, 1–18. Cambridge: Harvard University Press.

———, ed. 1999b. *Colonial modernity in Korea.* Cambridge: Harvard University Press.

Shin Yong-ha. 1995. Ku Chosŏn Ch'ongdok chŏngsanŭn harusoki chŏlgŏhaeya handa (The Government-General Building should be destroyed as soon as possible). *Wŏlgan Chosŏn* 178: 603–611.

Shinbu, Junpei. 1901. *Kanhantō* (The Korean Peninsula). Tokyo: Tōkyōdō.

Sibley, David. 1995. *Geographies of exclusion: Society and difference in the West.* London: Routledge.

Silhwa Kyeryongsan. 1965. (The real story of Kyeryong Mountain). Seoul: Han'guk Ilbo.

Sŏh Chŏng-Kwŏn. 1999. Nosukcha taechaek chahwal chungsim ŭro (Actualization of promoting homeless self-support). *Segye ilbo,* Seoul Section, September 14: 20.

Sŏk Pŏb-u. 1984. "Puch'ŏnim Maŭl" e taehayŏ (About "Buddha Village"). *Puch'ŏnim Maŭl* (Buddha Village) 2: 1–3.

Son Chŏng-mok. 1996. *Ilche kangchŏmgi tosi sahoesang yŏn'gu* (Facets of urban society during the Japanese occupation period). Seoul: Ilchisa.

Sŏn Tae-in. 1999. Changkwan nosukcha ŭi simya taehwa (Evening chat between the minister and homeless people). *Tong-ah Newspaper,* Social Section, July 22: 21.

Son Yul-lak. 1987. Tapsa Ilgi (Field study journal). *Ch'ŏn'go* (Antiquity) 53: 153–157.

Song Chae-jung. 1996. Namsan chido chejak (Producing the Namsan map). *Silla Munhwa* (Silla Culture) 10: 32–33.

Song Ch'ang-sŏk. 1999. Pom nosukcha ka tasi onda (Spring, homeless people come out to street again). *Han'gyŏreh* 21(251). Web site http://www.hani.co.kr/h21/data/L990322/1p9h3m0y.html.

Song, Jesook. 2006. Family breakdown and invisible homeless women: Neo-liberal governance during the Asian debt crisis in South Korea 1997–2001. *Positions: East Asia Cultures Critique* 14(1): 37–65.

Song Sang-gi. 1700. Yugyeryongsan'gi (Traveling through Kyeryong Mountain). *Ogojaejip* 13.

Song Sŏk-ha. 1934a. Minsokhakŭn muŏtinga? (What is folklore?). *Hakdŭng* (Study lamp) 4: 20–22; 5: 21–23; 6: 27–29.

———. 1934b. Ch'anggan sa (First editorial statement). In *Chosŏn minsok* (Chosŏn folklore) 1.

Sorensen, André. 2002. *The making of urban Japan: Cities and planning from Edo to the twenty-first century.* London: Routledge.

Spargo, Tamsin. 1999. *Foucault and queer theory.* Duxford, England: Icon Books.

Spencer, Joseph E. 1954. *Asia, east by south: A cultural geography.* New York: Wiley.

Stoklund, Bjarne. 1993. International exhibitions and the new museum concept in the latter half of the nineteenth century. *Ethnologia Scandinavica* 23: 87–113.

Sturdevant, Saundra Pollock, and Brenda Stoltzfus. 1992. *Let the good times roll: Prostitution and the U.S. military in Asia.* New York: New Press.

Sun, Soon-hwa. 1991. Women, religion, and power: A comparative study of Korean shamans and women ministers. Ph.D. diss. Drew University, Madison, NJ.

Sunagawa, Hideki. 2006a. Japan's gay history. *Intersections* 12. Web site http://wwwsshe.murdoch.edu.au/intersections/issue12/sunagawa.html. Accessed March 12, 2006.

———. 2006b. The social situation facing gays in Japan. *Intersections* 12. Web site http://wwwsshe.murdoch.edu.au/intersections/issue12/sunagawa1.html. Accessed March 12, 2006.

Taihi chūkai keisatsu hansho batsurei. 1920. (Comparative notes on the regulations for criminal punishment by the police). Tokyo: Jichikan henshūkyoku.

Takagi, J. T., and Hye-jung Park, directors. 1995. *The Women Outside.* Third World Newsreel.

Tanaka, Stefan. 1993. *Japan's Orient: Rendering pasts into history.* Berkeley: University of California Press.

Tangherlini, Timothy R. 1998. Shamans, students and the state: Politics and the enactment of culture in South Korea, 1987–1988. In *Nationalism and the construction of Korean identity,* ed. Hyung Il Pai and Timothy R. Tangherlini, 126–147. Korea Research Monograph 25. Berkeley: Institute of East Asian Studies, University of California.

Time Magazine. 2002. *Time Magazine.* August.

Times on the Web. 2003. Korean actor's reality drama: Coming out as a gay. *Times on the Web.* Accessed September 23, 2004.

Tonga Ilbo. 1923–1924. Kyeryongsan'gi (Reporting on Kyeryong mountain). *Tonga Ilbo.*

Trouillot, Michel-Rolph. 1995. *Silencing the past: Power and the production of history.* Boston: Beacon Press.

Tsukiashi, Tatsuhiko. 2000. "Hogokokuki" ni okeru Chōsen nashyonarizumu no keisei: Itō Hirobumi no kōshitsu riyōsaku tono kanren de (The formation of Korean nationalism during the "Protectorate period" as it relates to Itō Hirobumi's policy of using the royal family). *Chōsen bunka kenkyū* (Korean culture) 7: 73–75.

Turner, John Elliot, Vicki L. Hesli, Dong Suh Bark, and Hoon Yu. 1993. *Villages astir: Community development, tradition, and change in Korea.* Westport, CT: Praeger.

Tyner, James. 2002. Geographics of identity: The migrant experience of Filipinas in northeast Ohio. *Asia Pacific Viewpoint* 43(3): 311–326.

United Nations General Assembly. 1948. *Convention on the prevention and punishment of the crime of genocide.* New York: UN General Assembly.

Waldman, Amy. 2004. Heavy debt and drought drive India's farmers to desperation. *New York Times,* June 6: A1, A5.

Walraven, Boudewijn. 1995. Shamans and popular religion around 1900. In *Religions in Traditional Korea: Proceedings of the 1992 AKSE/SBS Symposium,* ed. Henrik H. Sorensen, 107–130. Copenhagen: Seminar for Buddhist Studies Monographs 3.

———. 1998. Interpretations and reinterpretations of popular religion in the last decades of the Chosŏn dynasty. In *Korean shamanism: Revivals, survivals, and change,* ed. Keith Howard, 55–72. Seoul: Royal Asiatic Society.

———. 1999. Popular religion in a Confucianized society. In *Culture and state in the late Chosŏn Korea,* ed. JaHyun Kim Haboush and Martina Deuchler, 160–198. Cambridge: Harvard University Press.

———. 2001. The parliament of histories: New religions, collective historiography, and the nation. *Korean Studies* 25(2): 157–178.

Watanabe, Manabu, and Hiroshi Abe, ed. 1988–1991. *Nihon shokuminchi kyōiku seisakushi shūsei, Chōsen-hen* (Collection of materials on the history of education policy in Japan's colonies, Korean volume), vol. 69. Tokyo: Ryūtani shosha.

Wells, Kenneth. 2003. Where from, who for? Modern Korean historical studies. *IIAS Newsletter* 32: 10.

Whang In-joung. 1981. *Management of rural change in Korea: The Saemaul Undong.* Seoul: Seoul National University Press.

Wilhelm, Richard, and Cary F. Baynes, trans. 1967. *The I Ching* (Book of changes). Princeton, NJ: Princeton University Press Edition.

Withers, Charles W. J. 1996. Place, memory, monument: Memorializing the past in contemporary Highland Scotland. *Ecumene* 3(3): 325–344.

Wolgan Chosŏn. 1995. *Wolgan Chosŏn* (Monthly Chosŏn) April: 389.

Yang Sŭng-u. 2003. Namch'on ŭi p'ilji chojik tŭksŏng kwa pyŏnhwa (Characteristics and changes in the structure of land plots in Seoul's "southern city"). In *Sŏul namch'on: Sigan, changso, saram* (Seoul's "southern city": Time, place, people), ed. Taehyŏng Hong. Seoul: Sŏulhak yŏnguso.

Yea, Sallie. 1994. Regionalism and political-economic differentiation in Korean development: Power maintenance and the state as hegemonic power bloc. *Korea Journal* 34(2): 5–29.

———. 2000a. Maps of the imaginary and geographies of dissent: The marginalisation of Cholla provinces in South Korea. Ph.D. diss. School of Geography and Environmental Science, Monash University, Melbourne, Australia.

———. 2000b. Maps of resistance and geographies of dissent in the Chŏlla region of South Korea. *Korean Studies* 24: 69–93.

———. 2002. Rewriting rebellion and mapping memory in South Korea: The (re)presentation of the 1980 Kwangju Uprising through Mangwol-dong Cemetery. *Urban Studies* 39(9): 1,551–1,572.

———. 2003. The politics of place: The role of the Kwangju Uprising in sustaining Cholla's democratic image. In *Contentious Kwangju: The May 18 Uprising in Korea's past and present,* ed. Gi-wook Shin and Kyung-Moon Hwang, 109–132. Boulder, CO: Rowman and Littlefield.

———. 2004a. *Report on sex trafficking of foreign women to United States military camp towns in South Korea.* Melbourne: Project Respect Publications.

———. 2004b. Runaway brides: Anxieties of identity amongst trafficked Filipinas in South Korea. *Singapore Journal of Tropical Geography* 25(2): 180–197.

———. 2006. Labour of love: Filipina entertainer's narratives of romance and relationships with G.I.s in U.S. military camptowns in Korea. *Women's Studies International Forum* 28(6): 456–472.

Yi Chae-gon. 1996. *Sŏul ŭi min'gan sinang* (Folk beliefs of Seoul). Seoul: Paeksan Ch'ulp'ansa.

Yi Chang-kon. 1998. Chŏsodŭk siljikcha wŏl 32 man wŏn kkaji chiwŏn. (Low income unemployed to be supported up to 320,000 won per month) *Han'gyŏreh sinmun,* April 17: 20.

Yi Ch'ŏng-jun. 1966. Pyŏngshin kwa mŏjŏri (Fool and idiot). *Ch'angjak kwa pip'yŏng* (autumn).

Yi Hae. 1712. Yugyeryongsan'gi (Traveling through Kyeryong Mountain). In *Kyŏngje sŏnsaeng yugo,* vol. 5. Unpublished manuscript.

Yi Hŭi-sŭng. 1972. *Kugŏ Taesajŏn* (Comprehensive dictionary of the national language). Seoul: Minjungsŏgwan.

Yi Kil-gu. 1996. *Kyeryongsan* (Kyeryong Mountain). Taejon Kwangyok-si: Taemunsa.

———. 1999. *Kyeryong sanmaek ŭn itta* (The Kyeryong Mountain Range). Taejŏn: Nuri Gihoek.

Yi Min-su. 1993. *Sinyŏk Chŏnggamnok* (The new translation of *Chŏnggamnok*). Seoul: Hongsinmunhwasa.

Yi Na-mi. 1999. *Uri ga sarang han namja* (Men whom we loved). Seoul: Haenaem Press.

Yi Nung-hwa. [1926] 1976. *Chosŏn musok ko* (Reflections on Korean shamanism). Trans. into modern Korean by Yi Chae-gon. Seoul: Paengnŭk.

Yi, T'ae-jin. 1999. Seoul at the beginning of the 20th century: Urban development based on Western models. *Korea Journal* 34(3): 95–121.

———. 2000. Chosŏn sidae "minbon" ŭisik ŭi p'yŏnch'an kwa 18 segi "minguk" inyŏm ŭi taedu (Changes in "people-centered" consciousness during the Chosŏn period and the eighteenth-century emergence of the theory of "unity between the king and his people"). In *Kukka inyŏm kwa taeoe insik: 17–19 segi* (The concept of state and foreign consciousness from the seventeenth to nineteenth centuries), ed. Ch'ung-sŏk Pak and Hiroshi Watanabe. Seoul: Ayŏn ch'ulpanbu.

Yi Wŏn-jae. 1998. *Han'gyŏreh sinmun,* Social Section, April 15: 23.

Yi Yong-sop. 1994. *Han'guk ŭi t'ŭksu pangmulgwan* (Korea's special museums). Seoul: Sesi.

Yoon Hong-key. 2001. Kyŏngbokgungkwa ku Chosŏn Ch'ongdok chŏngsanŭn ku Chosŏn Ch'ongdok kŏnmul kyŏngkwanŭl tulŏssan sanjingmul chŏnjaeng (The history of Kyŏngbok Palace and the iconographical warfare in the cultural landscape of Seoul). *Kongkankwa sahoe* 15: 282–305.

Yu Hong-june (Hong-jun). 1993. *Na ŭi munhwa yusan tapsagi 1* (The chronicle of my field study of cultural remains 1). Seoul: Ch'angjak kwa Pip'yongsa.

———. 1999. *Smiles of the baby Buddha: Appreciating the cultural heritage of Kyŏngju.* Trans. Charles M. Mueller. Seoul: Changbi Publishers.

Yun Hŭng-no. 1991. Chŏng In-sŏp. In *Han'guk minjok munhwa taebaekkwa sajŏn* (Encyclopedia of Korean culture), Han'guk minjok munhwa taebaekkwa sajŏn p'yŏnch'anbu, 19: 912. Sŏngnam: Han'guk Chŏngsin Munhwa Yon'guwon.

Yun Kyŏng-nyŏl. 1979a. *Kyŏngju Namsan kojŏk sullye* (Pilgrimage to the ancient relics of Kyŏngju Namsan). Kyŏngju: Kyŏngju-si.

———. 1979b. Kyŏngju Namsan yujŏk (Relics of Kyŏngju Namsan). *Ch'ŏn'go* (Antiquity) 50: 26–31.

———. 1993. *Kyŏre ŭi ttang, Puch'ŏnim ttang: Kyŏngju Namsan* (Land of our people, land of the Buddha: Kyŏngju Namsan). Seoul: Pulchisa.

———. 1994a. Kyŏngju pangmulgwan hakkyo 40 nyŏn paljach'ui (Footprints of 40 years of the Kyŏngju museum school). In *Hanŭl to nae kyosil ttang to nae kyosil* (Heaven too my classroom, earth too my classroom): ed. Kyŏngju National Museum, 79–83. Kyŏngju: Kyŏngju National Museum.

———. 1994b. *Namsan: Discovering the ancient wonder of Kyŏngju's Mt. Namsan.* Trans. Charles Mark Mueller. Seoul: Buddha Land.

———. 1997. *Majimak Sillain Yun Kyŏng-nyŏl* (Last person of Silla Yun Kyŏng-nyŏl). Seoul: Hakkojae.

Yun Sŭng-yong. 1994. Kyeryongsan chonggyoch'on ŭi minjungsinang (Popular beliefs in the religious villages on Kyeryong Mountain). In *Kyeryongsanji* (The chorography of Kyeryong Mountain). Ch'ungchŏngnamdo.

Contributors

Todd A. Henry is an assistant professor of history at Colorado State University. His teaching and research focus on the intersection of modern Japanese and Korean history, urban history, and the history of gender and sexuality.

Jong-Heon Jin is a lecturer at Seoul National University, where he teaches courses on human geography. His research on Korea has included studies of symbolic architectural landscapes, mountains, and Korean identity.

Laurel Kendall is curator of Asian Ethnographic Collections at the American Museum of Natural History and adjunct professor of anthropology at Columbia University, where she teaches in East Asian, popular religion, and museum studies.

David J. Nemeth is professor of geography and planning at the University of Toledo. He specializes in geography of Asia, cultural geography, and informal economies. His research and teaching interests include development planning on the Korean Peninsula.

Robert Oppenheim is an assistant professor in the Department of Asian Studies at the University of Texas at Austin. His research has centrally been on issues of place, cultural politics, and contemporary history.

Michael J. Pettid is an assistant professor at the State Unversity of New York at Binghamton, where he teaches Korean literature, language, and religions. His research includes studies on shamanic narratives, premodern literature, and women and Confucianism in Chosŏn Korea.

Je-Hun Ryu is a professor at the Korea National University of Education, where he teaches courses on cultural and historical geography. His current research interest is in the development of cultural and geographical theory in the study of Korea.

Jesook Song is an assistant professor at the University of Toronto, where she teaches anthropology of Korea and discourse analysis. Her current book project focuses on unemployment and homeless politics during the Asian debt crisis. She also conducts research on family-independent women in South Korea.

Timothy R. Tangherlini is a professor at the University of California, Los Angeles, where he teaches courses on Korean folklore. His research on Korea has included studies of shamanism, folk narrative, and popular culture.

Sallie Yea is a senior research fellow in international development at RMIT University, Melbourne. Her research interests include gender and development, human trafficking, and issues of tourism and identity and tourism and sustainable development.

Index

Production Notes for Tangherlini and Yea / *Sitings: Critical Approaches to Korean Geography*

Designed by University of Hawai'i Press production staff with New Baskerville text and display in Americana

Composition by Josie Herr

Printing and binding by The Maple-Vail Book Manufacturing Group

Printed on 55# Glat Offset B-18, 360 ppi